The Pocket Book of BACKYARD EXPERIMENTS

First published in the United States of America in 2020 by
Universe Publishing, A Division of
Rizzoli International Publications, Inc.
300 Park Avenue South
New York, NY 10010
www.rizzoliusa.com

Conceived, designed, and produced by
UniPress Books Limited

Designed by Lindsey Johns
Illustrated by Sarah Skeate
Project managed by Kate Duffy

Printed in China

2020 2021 2022 2023/ 10 9 8 7 6 5 4 3 2 1

ISBN: 978-0-7893-3803-7
Library of Congress Control Number: 2019952036

Visit us online:
Facebook.com/RizzoliNewYork
Twitter: @Rizzoli_Books
Instagram.com/RizzoliBooks
Pinterest.com/RizzoliBooks
Youtube.com/user/RizzoliNY
Issuu.com/Rizzoli

Discover the Laboratory in Your Garden

UNIVERSE

CONTENTS

WONDERFUL WILDLIFE

SOIL SCIENCE

FASCINATING FLORA

KITCHEN SINK SCIENCE

INTRODUCTION

If you've ever asked a question or wondered why something is the way it is, then you are a scientist. Scientists are interested in how the world works. They are inquisitive. They ask lots of questions. They make predictions, or "hypotheses," and then design experiments to test if their hypotheses are true. As new results come in, scientists are constantly revising what they think about the world.

This book is designed to bring out your inner scientist. It will help you to appreciate the natural world and the wonders that live in it. It's packed full of experiments and activities that will stretch your imagination and foster your natural curiosity.

All of the activities are centered around the backyard or garden. They can either be done in these

spaces or use items that come from them. If you don't have a backyard, don't worry. Head to a local park or a wild green space instead. Some of the activities can be done in small spaces, such as a balcony, window box, or plant pot, and some can be done indoors. All of them derive their inspiration from the natural world and the fascinating things that live in it.

Many of the experiments can safely be done without adult help. Where adult assistance is recommended, it is clearly marked at the start of the experiment. The activities in this book are designed for young people aged eleven or older, but younger kids will also enjoy them—they may just need a little more help.

The book is split into four different sections: wonderful wildlife, soil science, fascinating flora, and kitchen sink science. You can work through them in order, or dip in and out of the different sections when something catches your eye.

The ingredients for each activity are clearly listed. Most of these are simple items that can be readily found in the garden and the home. The instructions are clear and concise, and there are fun facts scattered through the book. Did you know, for example, that butterflies taste with their feet?

So, what are you waiting for? Get digging, planting, growing, making, spotting, designing, decorating, and experimenting. But most of all, get in your backyard and have some fun!

1

WONDERFUL WILDLIFE

It doesn't matter if you live in the middle of a big city or on a farm in the middle of the country—wildlife is everywhere. The earth is buzzing with things that fly, swim, crawl, hop, and run. All too often we overlook the creatures that live on our doorstep, but if you take the time to stop, look, and listen, you'll realize just how busy your backyard really is.

WONDERFUL WILDLIFE

Pollinators such as bees and butterflies (see the peacock butterfly, opposite) visit the blooms in our patio planters and flower beds. Amphibians hide in moist, dark places. Birds belt out tunes from high in the trees, and swoop down to nibble the treats found in bird feeders. Creepy-crawlies — such as snails, spiders, and beetles — patrol the borders, while caterpillars cling to their food plants and munch away quietly.

If you're lucky enough to have a pond nearby, you will see it is teeming with life. Ponds are wildlife magnets. They're a draw to water-dwelling creatures such as water striders and beetles, as well as frogs and damselflies (see below). Land animals such as birds and some mammals also come to drink from their edges.

It's great to look at and wonderful to study, but beyond that, the wildlife in our backyards plays a vital role. Every creature is part of an interconnected web of living creatures and their habitats, and they all have a role to play. By transferring pollen, pollinators help plants to reproduce. Birds help to keep the insect population in check. Caterpillars provide food for birds, and snails help to consume rotting plants and leaves.

It's time to celebrate the wildlife in your outdoor areas, and learn a little more about it. The activities and experiments in this section of the book are designed to make you take notice of the wildlife that surrounds you. There are experiments to help you attract and record wildlife, and understand the needs that different living things have.

Along the way, you'll learn how to give the wildlife in your space a helping hand, by putting out food and creating new habitats for wildlife to thrive. You'll be making homes for hoverflies, painting molasses on tree trunks to attract moths, and collecting caterpillars in an umbrella.

Best of all, once you start noticing and caring for the wildlife, you'll never stop. It doesn't take much to make your yard a wilder place, and welcoming wildlife is one of the most rewarding things you can do.

MAKE A SCIENCE JOURNAL

Scientists always write up their experiments so they have a permanent record of what they have learned. Create your own science journal so you can keep notes of all the activities you try from this book.

To make a science journal, find an unused notebook. One with a hard back is good because it's easier to write in when you're outside. If you don't have a notebook, you can staple some blank pieces of paper together.

Write your name on the front cover and label the notebook "Science Journal." Decorate it with doodles or cut out pictures from magazines and stick them on. Plants, animals, and landscapes would all look good, but this is your journal. Decorate it however you like.

Record every experiment and activity that you do. Start each one on a new page. Scientists always write up their experiments in a particular way, so this a good model to follow:

First, write the date and the title at the top of the page, and then a brief description of what you are going to do. This will be the Introduction section. For example:

MARK AND RECAPTURE SNAILS
An experiment to determine how many snails are living in a corner of the yard.

List all the items you are going to use, then write step-by-step instructions describing how you are going to do it. That way, if you ever need to repeat the experiment (good scientists do this a lot), you'll know exactly what to do. This is the Methods section.

When you've finished your experiment or activity, write up the results. If you've made observations or collected data, write it down. If you've built something, draw a diagram of it or take a photo and stick it in. This is the Results section.

Write up each experiment or activity in four sections with four different headings: Introduction, Methods, Results, and Discussion.

Now write down what you have learned. How successful was the activity? Did the thing you were building work as you'd hoped? How could it be improved? What can you learn from the results of your experiment? Look at the information that you've gathered, and draw a conclusion. This is the Discussion section.

BIRDSONG PLAYLIST

People often walk around wearing headphones, but when we're busy listening to music, we're missing out on one of the best sounds around: birdsong. Make a playlist of the birdsongs that you hear, and learn to recognize the calls of your favorite birds.

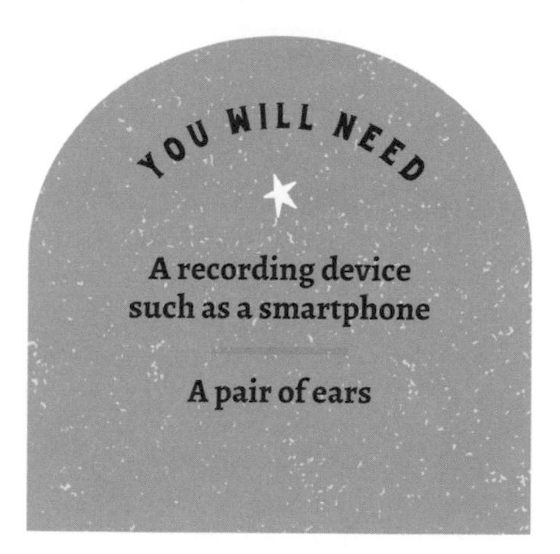

The best time to record birdsong is during the day when there are fewer birds singing and less background noise. Using your recording device, start by recording five different birdsongs. See if you can spot the birds that are making these noises. It's often easier to recognize a bird by sight than it is by sound, so this helps to match the bird to the song.

Different bird species sing different songs. Some belt out big melodies, while others sing more softly. Some of the tunes are long and complex, while others are short and simple. When you first start learning, it can be difficult to tell these different songs apart, so it's a good idea to record them and play them back.

When you are at home, try to identify the remaining songs. There are lots of online resources that can help with this. Concentrate on the structure of the song. Are there unusual features or parts that are repeated? Some songs contain recognizable sections that are unique to particular species.

Learning to identify birdsong is like learning a new language. The more you practice, the easier it gets. Label the songs you have identified and listen to them over and over. You can wear your headphones for this bit! If you'd like, you can reorder the tunes to create your very own birdsong playlist.

As you become familiar with different birdsongs, you can start to record where and when you hear them. Make a note of these observations in your science journal.

Birds make a remarkable variety of different noises. Birds can hiss, wheeze, growl, and even "boom."

Birds sing for lots of different reasons, such as to attract mates or defend their territories. Sometimes they change the sounds that they make. Some birds, for example, make high-pitched alarm calls to warn others that there are predators around. Try to recognize and record some of these more unusual calls. Do you notice any patterns emerging? Are some songs heard more often in particular places? If they are, you may have stumbled upon a bird that is defending its patch.

BIG BIRD COUNT

Scientists think that more than one in ten bird species are in danger of going extinct. Even common species are becoming rarer. Keep a record of the birds in your neighborhood by counting them to see how they are doing.

Find somewhere comfy to watch the birds. If you have an outdoor area where birds visit, you could choose to watch them through a window. Alternatively, find a chair to take outside and then settle down in a good spot. If you don't have an outdoor area, head to a park or a wild green space. Many birds live in cities, so if you find yourself surrounded by buildings, that's fine, too. Just have a look and see what you can spot.

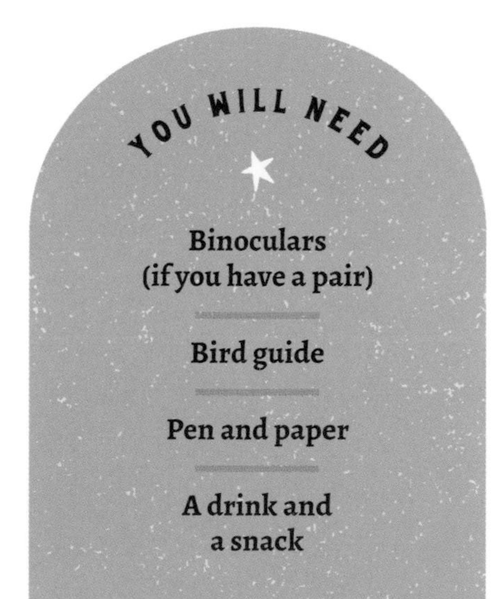

Sit quietly and don't make any sudden movements. Relax, have a drink and a snack, and watch the birds for an hour. Using your bird guide, make a note of all the different species that you see. Use obvious features, such as the bird's size, shape, and coloring, to help identify what it is.

Record your data because you never know when it will be useful. Some universities and wildlife charities run citizen science projects where they encourage members of the public to record their bird sightings and send them in. This is a great idea because it helps scientists to understand how bird populations are changing and how we can all help to save endangered birds.

As well as making a note of the species, record the number of individual birds that you see. Make a note of the maximum number of each species that you see at any one time. So, if you see a group of two sparrows together, and then later on a cluster of four sparrows, the number to write down is four. This means you are less likely to double-count the same birds.

Keep going back to your spot and see how the bird life changes over time. Some species, for example, are more numerous in the spring and summer because that is when they breed. Sometimes new species of birds appear and then disappear. These could be migratory species that have traveled vast distances. Their numbers may change across the day. Some species, for instance, are more active during the day and others are more active at night.

INSECT UMBRELLA

There are more than 150,000 species of butterflies and moths. Their caterpillars have evolved to feed on many different food plants. Determine the food preferences of tree-dwelling caterpillars with this simple insect umbrella experiment.

Most moths and butterflies lay their eggs when the weather warms up. The eggs hatch into caterpillars, so if you want to find them, this is an activity for a warm spring or summer day.

Head into your yard or a park or wild space and look for a deciduous tree with low-hanging branches. Deciduous trees are the ones that lose their leaves in winter. Insects find their leaves delicious! Avoid trees that are evergreen. Evergreen trees keep their leaves all year round. They tend to have glossy, waxy leaves that caterpillars don't like.

Once you have found your tree, start by looking at its leaves. If they look nibbled or have tiny holes in them, then something has been snacking on them. If the holes look fresh, the insect could be nearby. See if you can find some of these insects and identify them.

Next, look for a slender branch that is at roughly head height. Open the umbrella and turn it upside down, so the ribs and the handle are pointing up to the sky. Then put it on the ground underneath the branch.

Now take your stick and hit the branch. Give it a few short, sharp taps. Any caterpillars that are clinging to the leaves should fall into the upside-down umbrella. Take care not to damage the tree. Don't hit the leaves or any fresh shoots, as you may harm them.

Have a look at what you've got. Count the different varieties and identify what you have collected using a field guide. Then turn the umbrella over and let the insects go nearby. Now try again with a different type of tree. Are the sorts of insect that you find the same or different? Make a list of the caterpillars that you find on the various different food plants.

Some caterpillars are specialist feeders that only eat one type of plant. Others are generalists that have evolved to eat a variety of different plants. Which are more common in your backyard?

SWEET TREATS FOR MOTHS

Many people think that moths are boring and brown, but the truth is very different. Moths are often brightly colored, come in many different shapes and sizes, and play an important role in the natural world. You can attract them using this sweet moth treat.

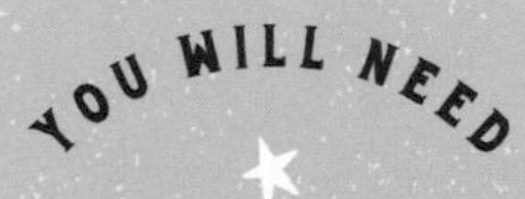

- **Bottle of cola 11 fl oz (330 ml)**
- **18 oz (500 g) dark brown sugar**
- **7 oz (200 g) molasses**
- **Large saucepan**
- **Paintbrush**
- **Flashlight**
- **Field guide**

This experiment works best during the summer. Although some moths fly during the day, most come out at night. Some of these are attracted to sugary solutions because they are similar to the nectar and sap that the moths usually feed on.

Prepare your sugary solution. Ask an adult to help you heat up the cola in a large saucepan. Let the liquid boil, then turn the heat down and let it bubble gently for five minutes. While it is simmering, add the sugar, one spoonful at a time. Keep stirring the liquid. When the sugar has dissolved, pour in the molasses. Then let it bubble for a few more

Although no one is really sure why, many moths are also attracted to light, so another way to attract them is to leave a bright outside light on during the night.

minutes. Finally, take the saucepan off the heat and set it to one side. Leave the mixture for a couple of hours, so it cools thoroughly.

At dusk, head outside with your mixture and your brush, and look for a place with lots of leafy, green trees. A yard or a park will do. Dip the brush into the mixture and paint it onto ten different tree trunks at eye-level height. Apply the mixture generously, creating big sticky patches on the tree bark. Warning—this can get messy! Then go home and wash your hands.

An hour or so later, when it is properly dark, go out and check the tree trunks. Shine the flashlight on the freshly painted bark. Photograph or draw what you find and add the pictures to your journal. Try to identify some of the species using a field guide, then repeat the experiment a few weeks later. Different moths visit the outdoor spaces at different times of year, so what do you find now? Don't worry about cleaning up. Next time it rains, the mixture will wash off.

BUG HOTELS

There are lots of bugs and creepy-crawlies that are good for the garden, but sometimes they struggle to find somewhere to live. It helps if your yard is not too tidy, but you can also build a deluxe bug hotel out of recycled materials and watch the creepy-crawlies move in.

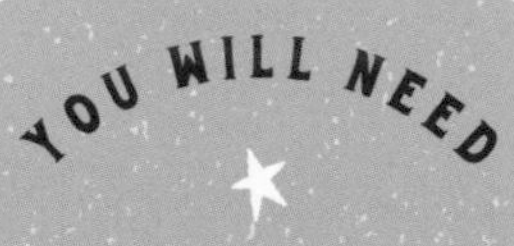

Some building materials such as:

Old pieces of wood

Straw, moss, dry leaves

Old plant pots, or patio tiles, bricks

Logs, pine cones, bamboo canes

Bark, woodchips, sand, or soil

Cardboard tubes or corrugated cardboard

Wood shingles, if available

First, think about where to put your bug hotel. You need a flat space to put it on, and a sheltered spot would be good. Different insects will move in depending on whether it's sunny or shady, so you may want to make two small hotels rather than one big one.

Start by making a sturdy structure. Think about how to best use your materials to make a hotel with several floors. Flat sheets of wood, supported on bricks, are an easy option.

Remember to put a roof on the top to keep out the rain. If you have wood shingles, then attach those on as the top layer. You may need a hammer and nails, and an adult to help. Asphalt roof shingles are another option.

Once you've built the outside of your hotel, it's time for some interior decorating! The idea is to create as many different nooks and crannies as you can, using different materials, to attract different bugs. You could stuff a section with straw or dry leaves, or cram it full of hollow stems such as bamboo canes. Can you fill a cardboard tube with moss, or roll up corrugated cardboard nice and tight? Some bugs like to burrow, so try filling an old plant pot with sand or soil and see what moves in.

You can make a very simple bug hotel by filling a terra-cotta plant pot with short sections of hollow stems (like bamboo). Lay it on its side somewhere that is sunny but sheltered, and wait for the bugs to arrive.

It will take a while for the first guests to arrive, but keep checking so that you can see what is living in your hotel. Can you name all the bugs and creepy-crawlies? Maybe you could draw pictures of them and keep them in a guestbook for your bug hotel.

MINI MAMMAL PRINTS

Record some of the small animals that visit your backyard by capturing their footprints in this mini mammal tunnel.

- Wash and dry the juice cartons, then cut the tops and bottoms off to make three long rectangular tubes. Slide the three tubes together to make one long tunnel. Now tape the tubes together.

- Line the bottom of the tunnel with a long thin sheet of clean white paper. Tape it in place.

- Check that the plastic lid will fit inside the tunnel. Now take it out and cut a piece of sponge so that it fits neatly inside the upturned lid.

- Moisten the sponge with water and add a good dash of food coloring. When you press your finger into the sponge; the coloring should stain your skin. If it's too faint, add a bit more food coloring.

- Fill the bottle top with peanut butter. You can use different types of bait, such as cheese or dog food, but peanut butter is good because it's sticky and prolongs the amount of time the mammal will spend standing in the food dye.

YOU WILL NEED

Three rectangular juice cartons

Scissors

Tape

White paper

Plastic container lid

Bottle top

Sponge (a makeup or dish sponge will do)

Food coloring

Peanut butter

Place the bottle top onto the sponge, and slide the plastic lid into the middle of the tunnel. Now carefully carry your construction into the yard. Rodents like to scurry along the sides of walls and fences, so place the tunnel next to a brick wall or garden fence. Camouflage the tunnel by covering it with leaves and branches.

DID YOU KNOW?
The front paws of mice have four toes, and their back paws have five toes. Rat trails are similar to mouse trails, but the footprints are larger.

Leave your tunnel for a few days. Check it every morning. Remove the plastic lid and look at the paper. If there's been a visitor, the footprints should be clearly visible on the paper. If you're lucky, you may even see a tail print, too! Stick the paper in your journal and see if you can identify the footprints. Make a list of all the different species you record.

Now try setting the trap in a wild space, such as a woodland. (Check first to see if you need permission.) There will be new species here. What do you find?

GROW PLANTS FOR WILDLIFE

Plants provide a source of food and shelter for all sorts of animals, such as insects and birds. Make your backyard more wildlife-friendly by adding a few well-chosen plants, then record the guests that come and visit.

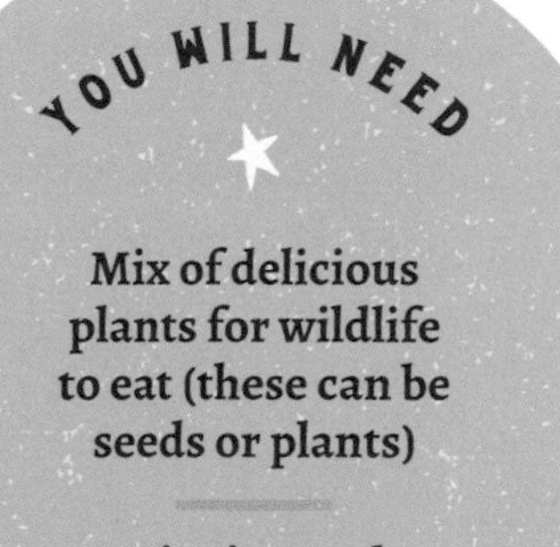

YOU WILL NEED

- Mix of delicious plants for wildlife to eat (these can be seeds or plants)
- Digging tool
- Water

Plants are the basis of all food chains. The more different plants you can grow, the more different kinds of wildlife you will attract to your garden.

Flowering plants are really important because they provide nectar and pollen for bees and other pollinators. Try to grow lots of different flowers. Go for simple plants with open flowers and single petals. Fussy flowers with multiple layers of petals may look pretty, but they often have less nectar and pollen.

Choose flowers that blossom at different times of the year, so you help the insects for as long as possible. Some insects like to stock up on food before they overwinter.

It doesn't take much to make your outside space more wildlife friendly, but if you don't have a garden, don't worry. You can still grow wildlife-attracting plants in pots on doorsteps, windowsills, and balconies.

Flowers are nice to have but don't forget about food plants. Many caterpillars and other insects graze on the leaves and stems of smaller plants, shrubs, and trees. Grow native plants because they are more likely to attract local insects than unfamiliar plants from far away.

As summer turns to fall, many birds and mammals eat the seeds and berries produced by some plants. Berries often contain a lot of fat, which helps to build the animals up before the arrival of winter, when there is less food around. Grow a mix of seed-bearing flowers, shrubs, and trees.

When you sow your seeds or dig in your plants, make sure you water them well. Then keep an eye out for them. New plants need a lot of watering while their roots get established.

If you have a garden, persuade your parents to let it get a bit messy. Fallen leaves help to fertilize the soil, and they provide a home for creepy-crawlies such as worms and beetles. Tangled climbers and bushy thickets provide a place for birds to build their nests.

Make a chart in your journal of the animals that visit the plants you have provided. Make a note of the date they visited and the plants that they used. Which is the best plant for wildlife?

BIRD BUFFET

It's always fun to watch birds feeding at a bird table or feeder, but have you ever wondered which snacks are their favorites? You can make a bird buffet and do an experiment to find out.

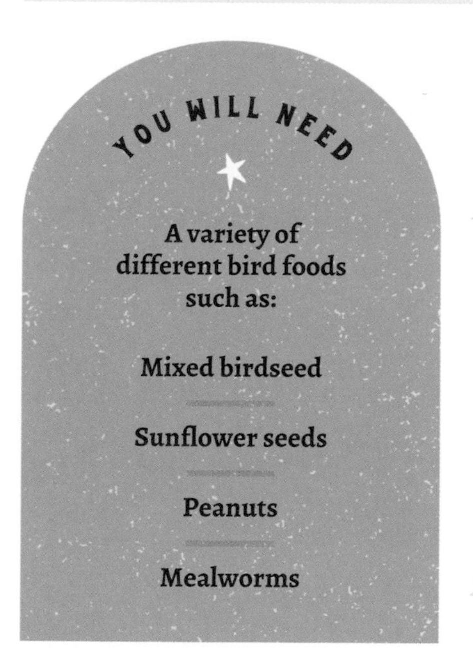

Lots of people like to leave food out for the birds. This is especially helpful in the winter when natural food is scarce, but it's also useful in spring and summer when parents are bringing up their young. In fact, if you feed the birds, it's a good idea to leave food out all year round.

Just like us, different birds like to eat different things. You can test their preferences by leaving out a variety of different bird foods. Mixed birdseed, sunflower seeds, peanuts, and mealworms are good foods to try, but you could also test them on other things like banana or grated mild cheese.

Choose the location for your feeding experiment. It should be somewhere safe, where the birds won't get ambushed by cats or other predators. Bird tables are good because they're high off the ground, but you could also set up the experiment on a balcony or brick wall.

Now lay out your bird buffet. Put down a handful of each food type. Leave a good space between each pile so the foods don't get mixed up.

If you want, you can set up the experiment using bird feeders. Fill different feeders with different ingredients and hang them up next to one another.

Now sit back and watch what happens. Don't be put off if the birds don't come immediately. Birds are often wary of unfamiliar situations, but once they realize the experiment is delicious and safe, they'll keep coming back!

Set up a chart with columns for the different bird foods, then add in the various species that eat them. Can you spot any patterns? Birds definitely have favorites. What do your birds enjoy the most? Now see if you can find something that they like to eat even more!

DID YOU KNOW?
British great tits are evolving longer beaks than their European counterparts. Scientists think this is because British people love to feed the birds, and the birds' beaks are adapting to the feeders.

HOVERFLY HAVEN

Although hoverflies look a bit like bees and wasps, they are not closely related. You can welcome these harmless pollinators into your backyard by building a hoverfly haven.

As their name suggests, hoverflies can often be seen hovering above colorful flowers. Hoverfly larvae are the gardener's friend since they often eat aphids and other troublesome pests. Many hoverflies are mimics. They have evolved to look like bees and wasps, but don't worry—they don't sting at all. This is a disguise that helps to protect them from predators.

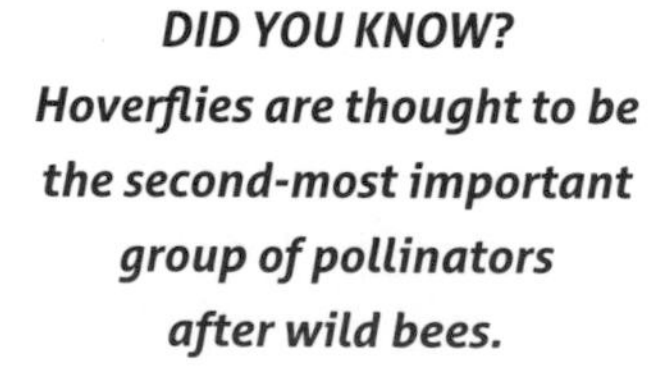

DID YOU KNOW?
Hoverflies are thought to be the second-most important group of pollinators after wild bees.

There are lots of different kinds of hoverfly. Some lay their eggs on plants, but some breed in still pools of water where their larvae eat decaying plant and animal matter.

Fill the ice cream tub halfway with grass clippings then pour in some water. Add enough water to cover the grass.

Next, add a generous handful of leaves. The leaves will rest on the surface of the water. This gives the female hoverflies somewhere to land when they come to lay their eggs.

Now add the sticks. The sticks need to poke out of the water. You can achieve this by sliding them into the grassy water and then leaning the sticks against the side of the ice cream tub.

Once the eggs have hatched, the larvae will live in the water. The larvae of some hoverfly species breathe underwater by poking a hollow tail-like tube out of the water, just like a snorkel! After turning into pupae, the adult hoverflies emerge. The sticks help them to climb out of the water, and give them a place to flex their wings before taking off.

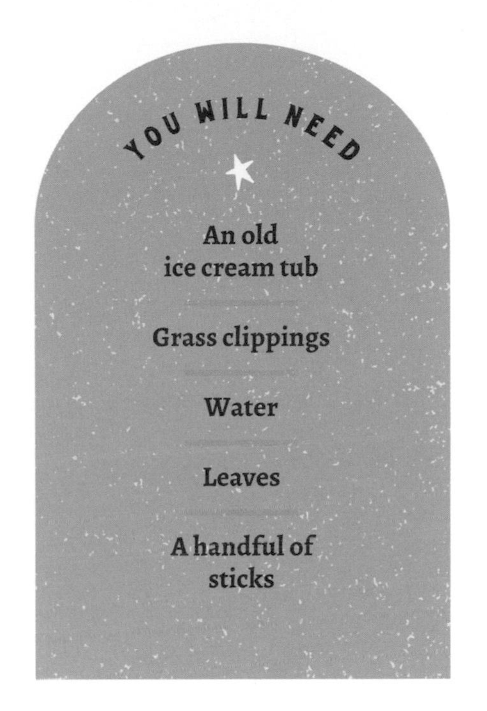

YOU WILL NEED

An old ice cream tub

Grass clippings

Water

Leaves

A handful of sticks

Check your hoverfly haven regularly. In warm weather, the water will evaporate, so keep topping it up. As the grass and leaves begin to break down, it will create the perfect conditions for your very own hoverfly nursery.

Photograph the hoverflies that you find around the haven. There are thousands of different species of hoverflies. Using an online guide, try to identify them. Stick the photos into your journal and see if you can work out which is the most common species in your patch.

ANT FARM

Ants are very industrious. They live in colonies and work together, helping to keep the environment clean by decomposing plant and animal matter. Record how long it takes for ants to create their home by making an ant farm in a jar.

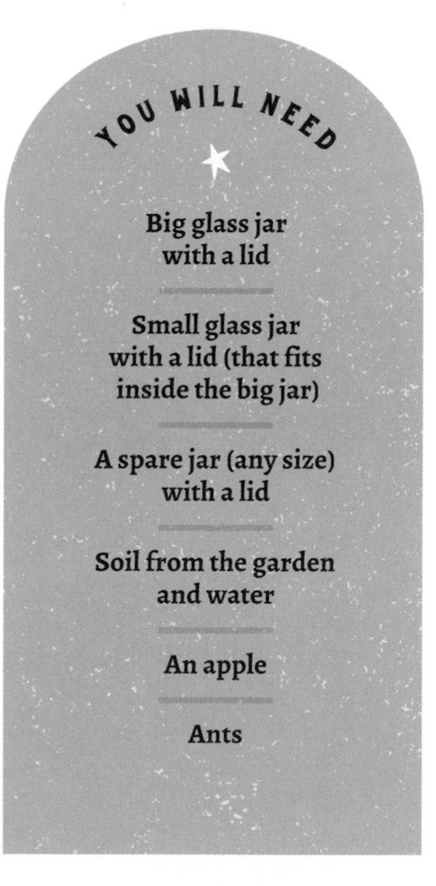

Take the labels off the jars, then wash and dry them thoroughly. With its lid on, place the smaller jar upside down inside the bigger jar. Make it as centered as possible. Start filling the big jar with dryish soil from the garden. Pour it down the sides of the big jar so the soil surrounds the little jar. Don't

completely cover the little jar. Leave the top of it poking out of the soil. Topsoil is good because it doesn't have any big lumps in it and is often quite sandy. This makes it easier for the ants to move around.

Now go outside and find some ants. Choose your ants wisely. Some varieties have a nasty bite, while others are harmless. Check with an adult to make sure the ants are suitable.

You will need to set an ant trap. Mush up a slice of apple with some water and put it into the spare jar. Then lay the open jar on its side on the ground and wait for the ants to wander in. They will be attracted to the sugar. You don't need too many ants—ten to twenty is perfect. Then close the lid and carry them to the ant farm.

DID YOU KNOW?
Some ants can lift more than twenty times their own body weight. If an eleven-year-old girl were that strong, she would be able to lift a cow!

Tip the ants and mushed apple into the ant farm, then quickly screw the lid on. Don't worry about making air holes. Ants are tiny and don't need much oxygen, so there will be more than enough inside the jar to keep them going.

Place the jar on a shelf or table—anywhere that is not in direct sunlight—and enjoy! The ants will move the soil around and create an amazing network of subterranean tunnels.

The ants will need feeding every few days. Swap the old apple for a new piece and add a tiny sprinkle of water. Do this outside in case any ants escape.

Record how long it takes for the ants to make their tunnels. Draw a map of the network they create and stick it in your journal.

WHAT DO WOODBUGS LIKE?

All living things have a preferred environment or "niche." Some like it hot. Some like it cold. Some like it dry. Others prefer it wet. Woodbugs are no different. Find out where they like to live using this simple experiment.

YOU WILL NEED

- Jar
- Small paintbrush
- Piece of paper
- Two shoeboxes with lids
- Ruler, scissors, and tape
- Dry and wet soil

Woodbugs are small crustaceans. They have a hard outer exoskeleton made up of separate segments and seven pairs of jointed legs. They can be found hiding under outdoor plant pots, stones, rotting logs, and leaf litter. Fill your jar with a small amount of soil and go out and collect twenty woodbugs. Use a paintbrush to flick the woodbug onto a piece of paper, then tip them into the jar.

Prepare the experiment. First, cut a rectangular hole measuring roughly 4 in × 1/2 in (10 cm × 1 cm) into one of the short sides, or "walls," of the shoebox. The hole should be right at the bottom of the wall, next to the base of the

shoebox. Now, cut a matching hole into the second shoebox. Check that it is the same size as the first hole.

Using some tape, join the two shoeboxes together, so the holes are perfectly lined up. You should now have one very long box with a wall that has a hole in it, in the middle.

Woodbugs aren't fussy eaters. They eat rotting plants, fungi, and their own feces, and they don't pee! They get rid of some waste products by excreting a pungent chemical called ammonia through their exoskeleton.

Place dry soil in one side, and wet soil in the other. Now add ten woodbugs to each compartment and put the lids back on the shoeboxes. The woodbugs will start to move around and may choose to move from one box into the other. Check the woodbugs every fifteen minutes and record how many there are in each compartment each time you look. Where did the woodbugs like to go and how long did it take them?

Now repeat the experiment, but this time, leave the lids off. Did the woodbugs still make the same choice? How long did it take them this time? Can you work out how fast they moved?

MAKE A BUG SUCKER

Backyards and open spaces are full of tiny bugs so small that they're difficult to pick up and study. Fear not! Suck them up with this homemade bug sucker or "pooter," and record how many different types of tiny bug are in your backyard.

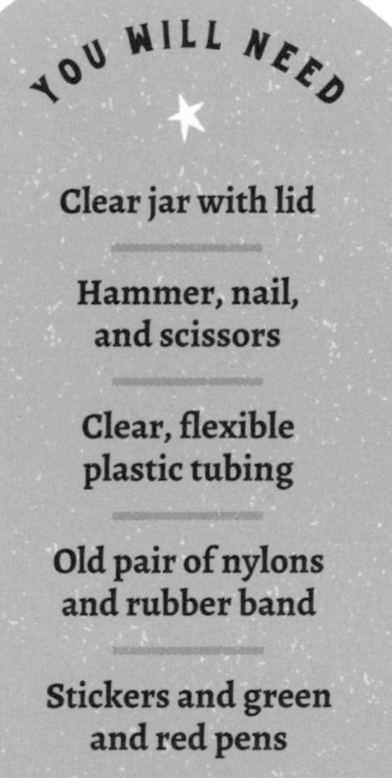

Using the hammer and nail, create two holes in the metal lid of the glass jar. Enlarge the holes so the plastic tubing will fit through them. You may need an adult to help with this.

Prepare the tubing. Clear, plastic tubing can be bought from a hardware store. You'll need about 20 in (50 cm). It should be 1/4–3/8 in (7–10 mm) wide. Cut the tubing into two pieces, about 6 in (15 cm) and 13 1/2 in (35 cm) long. Poke the tubing through the holes in the lid. There should be about 2 in (5 cm) of tubing on the inside of the jar once everything is assembled.

- Cut a circle from the pair of nylons. It should be about 2 in (5 cm) across. Wrap the stretchy fabric around the end of the shorter tube, then fasten it in place using the rubber band. When the jar is reassembled, this should be on the inside of the jar.

- Color one of the stickers green, and the other one red. Stick the green sticker onto the other end of the short tube. This is the tube you will suck through. Mark the end of the longer tube with the red sticker. This is the nozzle of the sucker.

- Make the lid airtight by sticking some putty or Plasticine around the joins where the tubes meet the lid. Screw the lid back onto the jar. Your insect sucker is now ready to go!

- Go outside and find some creepy-crawlies. Aim the nozzle end of the sucker (with the red sticker) at the insect you would like to collect, then give a short, sharp suck on the end of the other tube (with the green sticker). The insect will be sucked up the nozzle and into the jar. You can now observe it carefully. Don't forget to let them go when you have finished.

How many different types of tiny insects can you find? Draw what you find in your journal. Look in various habitats, such as dark, shady places and exposed, sunny flower beds. Do you find different insects in other habitats?

TOAD SHACK

Every creature deserves somewhere nice to live, and toads are no exception. See if you can attract them to your garden by building them their very own toad shack.

Paint a colorful design or picture on the plant pot. You could paint something that a toad might like, such as lots of delicious flies.

YOU WILL NEED

- Medium plant pot
- Paints and a brush
- Digging tool
- Leaves
- Shallow bowl

Toads are amphibians. They have thin, permeable skin. This means that liquids and gases can pass through the skin easily. This is good, because it helps them to breathe—and bad, because they dry out easily. That's why toads like to live in damp, shady places. Choose a location for your toad shack. It should be somewhere shady with lots of plants and places to hide. This will help to keep the toad's skin moist. Flower beds and overgrown corners are good spots.

Dig a hole that is big enough to fit the plant pot when it's on its side. Don't make it too deep. The top half of the plant pot should be above ground.

Turn the plant pot on its side and place it in the hole. Add a handful of wet leaves so the toad has somewhere nice to sit.

You might think that toads need to live in ponds. Sometimes they do, but they will also be quite happy in the toad shack as long as there is a source of water nearby. Fill the shallow bowl with water and place it near the toad shack.

Be patient. It could take weeks or even months for a toad to move in. Keep checking the shack on a regular basis. You can also look for toads in dark places under logs and leaves. Toads have bumpy, rough skin and frogs have smooth skin that is covered in mucus. When you do find a toad, pick it up carefully and put it in its new home. Now wash your hands. It's up to the toad to decide if it wants to stay.

DID YOU KNOW?
Toads catch beetles, spiders, and other prey using their long sticky tongues. Unlike human tongues, which are attached at the back of the mouth, toad tongues are attached at the front of the mouth. This is because toads don't chew their food. They just swallow it straight down. Gulp!

POND DIPPING

Ponds are absolute havens for wildlife. Beneath the surface of the water, there's a whole world of weird and wonderful creatures just waiting to be found. Find out what's lurking in the depths of your local pond.

YOU WILL NEED

Shallow tray or big ice cream tub

Net (see pages 160–61 for how to make one)

Field guide

Magnifying glass (if you have one)

Smartphone with camera (if you have one)

Science journal

Jam jar

Fill the tray halfway with water from a pond and put it to one side on level ground. Avoid placing the tray in direct sunlight as this will warm the water and could harm the animals that you catch.

Dip the net into the pond and sweep it through the water in a figure-eight shape. This helps to make sure that anything you catch will stay in the net.

Don't lean too far out. The areas around the edge of the pond will have lots of life in them. Be careful not to scoop up mud or stir up the sediment at the bottom of the pond. This will clog up your net and make it difficult to see what you have caught.

Take a trip to your local pond. If you're lucky, you might have a pond in your garden. If you don't, head out to a local park or nature reserve and see what's hiding in their waters.

Empty the net by carefully turning it inside out onto the tray. Let the water settle. Take a look at what you have caught and use your field guide to identify the different creatures. Some of the animals will be very small, so use the magnifying glass to help you. If you're struggling to identify something, take a photo, then look it up later. Make a note of all the different creatures that you find in your science journal.

Don't keep the animals in the tray for too long. Once you have finished with them, use the glass jar to scoop them up and return them to the pond. Keep searching. When you have completely finished, tip the contents of the tray back into the pond. Make sure you wash your hands after this activity.

A BIRDBATH FOR ALL SEASONS

As their natural habitat disappears to make way for towns, cities, and agricultural land, birds are becoming increasingly reliant on gardens. Help them by making them a birdbath, then record all the visitors that use it.

YOU WILL NEED

- Shallow tray (a metallic trash can lid is ideal)
- Three bricks
- Water
- A big stone and some gravel
- Tea light candles and matches

Choose your dish. The birdbath will be out in the garden all year round, so it needs to be made of something, such as metal or ceramic, that won't shatter when it freezes. A large, metal trash can lid makes an excellent birdbath because it has shallow, sloping sides and is nice and wide. Plastic alternatives won't work because in the winter the birdbath will be heated, and the tray could melt.

Find somewhere flat to put the birdbath. It should be easily visible, and ideally, in a big open space with trees and bushes nearby. This will help the birds keep an eye out for predators while they visit the birdbath. If cats come to your

Birds need fresh water all year round for drinking and bathing. This is especially important in the winter, when water freezes, and in the summer, when it evaporates more easily.

garden, prune back any trees or bushes that are within pouncing distance. This will make it harder for the cats to ambush the birds.

- Lay out three bricks on the ground in a triangular shape, with a space in the middle. Balance the tray on the bricks, making sure it is stable. Add a large stone to the middle of the tray, surrounded by a sprinkling of small stones or gravel. The stones provide a perch for birds to sit on, and a lifeline for any beetles or bugs that fall in—they'll be able to crawl onto the stones, dry themselves, and then fly away.

- Fill the birdbath with water. It should be no more than 4 in (10 cm) deep. In the winter, when the temperature drops, place a lit tea light candle in the middle of the bricks underneath the birdbath. This will help to prevent the water from freezing. Check the birdbath every day and fill up the water as needed.

- In your journal, record the different birds that use the bath. Note the date when they visited and what the weather conditions were like. Do the birds use the bath more at certain times of year?

MAKE A BUTTERFLY FEEDER

Butterflies are welcome visitors to any outdoor space. These important pollinators always add a splash of color. Find out how to encourage them into your garden by making a vibrant butterfly feeder.

YOU WILL NEED

- Paper cup
- Sharp pencil, scissors, and string
- Kitchen sponge
- Colored paper and glue
- Apple juice
- Field guide

Using the pencil, make two holes on opposite sides of the paper cup. The holes should be just below the rim of the cup. Cut a piece of string about 15½ in (40 cm) long. Poke the ends through the holes in the cup. Tie a knot at each end so the string makes a handle.

Using the pencil again, make a hole in the center of the bottom of the cup. Use the pencil to widen the hole. The hole should be a bit bigger than the width of the pencil. Cut a piece of sponge that is at least twice as big as the hole. Now push the sponge into the hole so half of it is sticking out.

Butterflies feed through their proboscis but taste through their feet. This means they can taste their food just by standing on it.

It's time to decorate the cup. Butterflies are attracted to brightly colored flowers, so cut out some paper flowers from the colored paper. Make any design you like, then stick the flowers to the side of the cup.

Fill the cup halfway with apple juice and hang it somewhere sunny in the garden, then watch and wait. Observe the butterflies as they come to feed. Butterflies feed using a hollow tube called a proboscis, which acts like a straw. When the butterfly is not feeding, the proboscis is curled up like a deflated party blower. When the butterfly is eating, the proboscis is unfurled. The butterfly below is called a painted lady. It's well known throughout much of the world.

Make a list of the butterflies that come to feed, and use a field guide to help identify them. You can experiment by swapping the apple juice for other flavored juices or for sugar water. (Sugar water is one part sugar and ten parts water.) Do certain butterflies prefer certain flavors?

MARK AND RECAPTURE SNAILS

Many gardeners dislike snails because they eat crops and flowers, but snails play a vital role in our gardens, where they consume rotting leaves and other decaying plant matter. See what they get up to by marking and recapturing them.

There are many different species of snails. Some live on land and some live in the water. This experiment focuses on land-dwelling snails. This is an activity for spring, summer, or fall, because this is when snails are most active.

Choose an area to hunt for snails. Snails avoid being out in the open too much as this makes them vulnerable to drying out and to predators that may spot them and gobble them up. Look in nooks, crannies, and other sheltered places. Snails can often be found under the rims of plant pots, on the shady sides of walls, and in the leaf litter on the ground.

Find as many snails as you can in your chosen area. Pick them up carefully using your forefinger and thumb, and transfer them into the box or container. Take care not to

YOU WILL NEED

- **Empty box or container**
- **Nontoxic, waterproof paint**
- **Small paintbrush**

Suppose you found ten snails on the first search. On the second search you found ten, and two of these were marked. The total number of snails in the area is 10 × 10/2 = 50.

leave the box in bright sunshine as this could harm the snails. Make a note of the number of snails that you find.

Using the brush, place a tiny dot of paint onto the shell of each snail. Make sure to use non-toxic, waterproof paint that won't wash off in the rain and leave the snail unharmed. Take care not to get paint on the snails' bodies as they are very sensitive. Then release the snails back where you found them.

A few days later, go out and do another search. Make a note of how many new, non marked snails you find and how many paint-spotted individuals you manage to recapture.

You can work out how many snails are living in your area by multiplying the total number of snails from the first search by the total number of snails from the second search, and then dividing the answer by the number of marked snails from the second search.

BUILD A DEAD HEDGE

Hedges that are made out of degrading plant material are called dead hedges. They're easy to make and great for wildlife. Build your own and find out what sorts of wildlife will move in.

Decide where you are going to make your dead hedge. Dead hedges are substantial structures that take up a bit of space, so you may want to check with an adult first. You can use them to create boundaries or to fence off key areas such as ponds and wildflower meadows.

Normal hedges can take decades to grow, but dead hedges can be made in an afternoon. With a little maintenance, they can last for years.

Ask an adult to help you drive the posts into the ground with the mallet so they are standing upright. The posts should be about 1½–3 in (4–8 cm) in diameter. You can use tree branches or posts from a garden center. It helps if the posts are sharpened at one end.

Arrange the first three posts so they form a row. Each post should be separated by a gap of about 20 in (50 cm). Tie a piece of garden string to one of the posts and then wind it back and forth between the posts so that it gives the boundary of the hedge some structure.

Make an identical second row of posts. There should be a 20 in (50 cm) gap between the two rows.

If you have any cuttings from stringy climbers, such as ivy, or bendy trees like willow, you can weave them in and out of the posts to reinforce the string.

★ Fill the gap between the two rows of posts with sticks, hedge cuttings, and deadwood. Anything woody that takes a while to rot will do. Pile the items on top of one another until they reach the top of the posts. Press it down gently to compact it. Fill up if necessary.

★ Don't add green waste such as grass clippings or leaf mulch. This will rot too quickly and turn your dead hedge into a compost heap!

YOU WILL NEED

Six posts or stout sticks measuring about 3 ft (1 m) tall

Mallet

Garden string

Twigs, branches, and other woody cuttings

★ Sit back and watch the wildlife move in. Dead hedges provide homes for birds and small mammals, not to mention invertebrates such as beetles and bees. Keep a record of what comes to visit, and compare what you find with the wildlife in and around a normal green hedge. Is there more or less wildlife in your dead hedge?

MAKE A SOUND MAP

Sound maps are simply maps of the sounds that surround us. There are lots of interesting sounds in the great outdoors. Take time to hear and record them by making your own sound map.

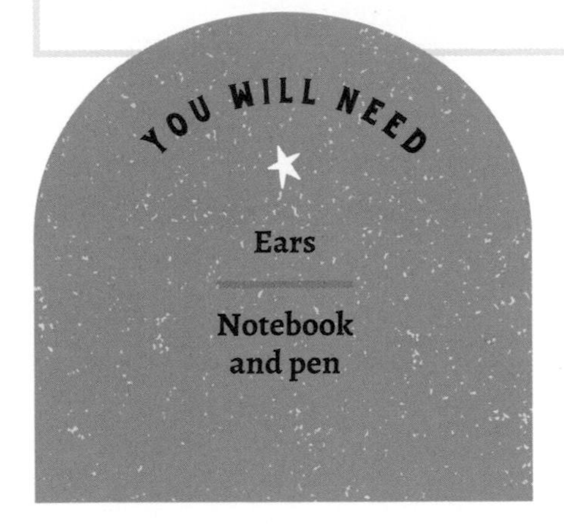

We rarely take the time to stand still and listen to the noises that surround us, but it's an experience that can be deeply rewarding.

Head outside, find a spot, and stand still. What can you hear? Sometimes it helps to close your eyes so you can concentrate on sounds rather than sights. At first, you'll notice things that are familiar. It might be birdsong, the barking of a dog, or the thrum of a neighboring lawn mower.

Be still and silent. Now listen more deeply. What else can you hear? When you really stop to listen, you can hear all sorts of sounds. You may hear the rustling of wind in the trees, the buzzing of bees, or the whining of tiny flies as they zoom past you. Every landscape has its own unique soundscape.

Write a list of all the sounds that you hear in ten minutes. Now draw your map. Draw a picture of yourself in the center of the map, then position all the other sounds around you.

Maps show us where things are and give us an idea of the distance between different objects. Try to make your map do the same. If a car sounded farther away than a dog, for example, draw it farther out on the map. If the dog was louder than the car, you may wish to make the dog bigger than the car.

Go back outside later in the day and repeat the process. How have the sounds changed? Are there any new noises? Maybe next time ask a friend to join you then you can compare the sound maps you both make. People often tune in to different sounds in their environment, so their map may be different from yours.

Make a sound map when you go somewhere different, such as to the seaside or the zoo. Compare the sounds from home with these new, exotic noises.

MAKE A HIBERNACULUM

A hibernaculum is a shelter that is used by animals to help them survive the cold. See if you can attract amphibians to your garden by building them a place where they can spend the winter.

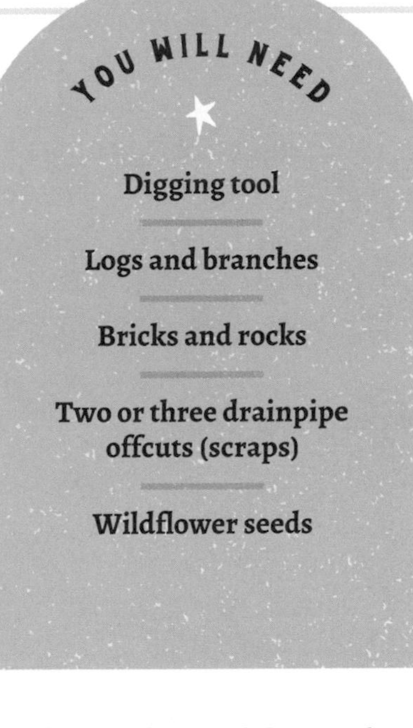

Find a spot for your hibernaculum. It's best to build it tucked out of the way, in a corner of the garden. You're about to create a very big hole, so make sure you're not about to uproot any important plants. Check with an adult to make sure the location is a suitable one.

Dig a hole that is about 60 in (150 cm) wide and 20 in (50 cm) deep. Pile up the earth you remove on the side of the hole. Fill the hole with logs, branches, bricks, and rocks. They can be piled in messily but the idea is to leave lots of space between these items so there are cavities where the animals can hide.

Now for the drainpipe scraps. If offcuts are difficult to come by, you can ask a builder or hardware store if they have some spare. The scraps should be roughly 12–20 in (30–50 cm) long. The drainpipes are going to be tunnels that connect the cold outside world with the warm, cozy hiding place.

Push the drainpipes into the hole so they nestle between the rocks and logs. One end of the drainpipe should be poking out above ground so it is clearly visible. The drainpipes should be slanted at an angle so the residents can walk, hop, or slither up and down the tunnels.

Cover it all up with the soil that you removed, but make sure the entrances to the drainpipes are left open. Build up the soil into a mound. This will give the hibernaculum extra insulation. Sprinkle the mound with wildflower seeds and water them in.

DID YOU KNOW?
Lots of animals hibernate. It's a way to save energy in the cold. Bears build their own hibernacula inside caves, hillsides, and gnarled old tree roots.

Your hibernaculum is now good to go. These wintery hiding places attract small animals, such as toads, lizards, snakes, and insects. Monitor the hibernaculum and keep a record of what goes in and what comes out. But try not to disturb the creatures. List and draw the different species in your science journal.

WATER-BEAR HUNT

Water bears, or tardigrades, are tough, tiny creatures that can live just about anywhere, from the deep ocean to mountaintops, from the tropics to Antarctica. They're enchanting to look at, so go on a water-bear hunt and see what you can find.

Go out and look for moss and lichen. Moss is often green and velvety. It tends to grow in clumps in damp and shady places. Lichens are often a green or blue-gray color. They tend to be crusty and can be found growing on exposed surfaces. You may find lichen on stones, tree trunks, sidewalks, or on the sides of buildings. Using the pocket knife, scrape lots of moss and lichen into your clean plastic box and bring the container home.

DID YOU KNOW?
Water bears are so tough, they can even survive the vacuum of space.

Water bears live in moss and lichen, but if it's too dry, the tiny creatures dry out and enter a state of suspended animation called cryptobiosis. Water reanimates them. Fill the container with rainwater and leave it overnight. If you don't have rainwater, use bottled spring water. Tap water contains chemicals that can harm the water bears.

The next day, scoop up the moss and lichen and squeeze it so all the water goes back into the plastic tub. Then pour a little of the water into a clean bowl or tray.

Water bears are so tiny that you'll need a magnifying device to see them. A microscope is ideal, but many smartphones now have a built-in magnifying function. Alternatively, you can buy a clip-on microscope that fits over the lens of a smartphone.

There will probably be lots of living things swimming about, so you need to know what you're looking for. Water bears have wrinkly bodies, scrunched-up heads, and four pairs of legs. They look like little bears, but some people call them moss piglets because they look a bit pig like too.

When you spot one, increase the magnification and take a photo for your journal. How many water bears can you see? Do samples from different places contain more or fewer water bears? Remember to set your water bears free in the wild when you are done.

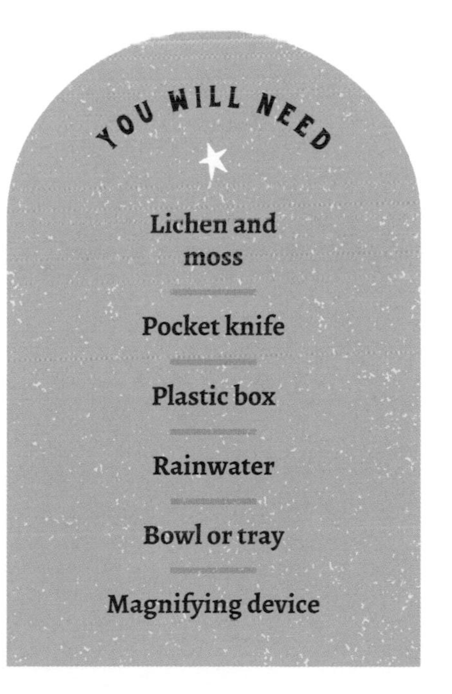

MAKE A BIRD FEEDER

Food shortages can occur at any time of year, so give our feathered friends a helping hand by hanging up some homemade bird feeders. Test different designs to see which is more popular.

Different birds like different foods, so some bird feeders may be more popular than others. We're going to make two different types of feeders and then test them in the backyard.

Which of your bird feeders is more popular? Do certain types of bird prefer one to the other? What times of day do the birds visit? Keep a note of your observations in your science journal.

For the first feeder, add the dry ingredients into a bowl. Garden birds will happily eat birdseed, but they'll also eat human food, such as bacon rinds, raisins, bread crumbs, and grated cheese. Mix the ingredients together.

Soften the solid cooking fat by leaving it on a windowsill or a heating vent. Add the cooking fat into the bowl and use a wooden spoon to mush all the ingredients together. The cooking fat will bind the ingredients together.

Take a long piece of string and tie one end around the handle of the mug. Fill up the mug with the bird food mixture and push the garden twig into the hole, so it is half in,

YOU WILL NEED

FEEDER 1

Dry ingredients such as birdseed, grated cheese, and bread crumbs

Hard cooking fat such as lard

Wooden spoon, string, and twig

Bowl and old mug

FEEDER 2

Apple, and sunflower seeds

Two sticks about 6 in (15 cm) long

String and scissors

half out. Now put the mug in the fridge so the fat becomes hard and sets around the twig.

While it is setting, prepare the second bird feeder. Ask an adult to core an apple by cutting a hole through the middle of the apple to remove the core. Take a handful of sunflower seeds and push them into the fleshy part of the apple so the seeds are half sticking out.

Take two short sticks and cross them so they form an X shape. Take a long piece of string and tie the crossed-over sticks together. Thread the long end of the string up through the apple core. The sticks make perches for the birds to stand on while they are eating the apple and the seeds.

Hang both feeders in the backyard and wait for the birds to appear.

MAKE AN UNDERWATER VIEWER

Check out the underwater wildlife in a nearby pond or stream. Make an underwater viewing device called a bathyscope and record the aquatic life-forms that live near you.

- Cut the top and bottom off your plastic bottle to make a long hollow tube. You may need to ask an adult to help with this. Use the sandpaper to file down the freshly cut surfaces so there are no sharp edges.

- Stretch the clear plastic over one end of the bathyscope. It's good to recycle, so you could use an old, clear plastic bag, or you could use the clear plastic wrapping that is sometimes used to protect fresh fruit and vegetables. Secure the plastic in place using a rubber band. Make sure it's stretched tight so you can see through it clearly.

- The bathyscope needs to be watertight, so once the plastic is in place, wrap duct tape around the outside of the bottle so the plastic becomes stuck to the side. This should prevent any water from leaking into the bathyscope when you use it.

Find a local pond or stream. It's time to test the bathyscope out. Pond dipping is all well and good, but a bathyscope enables you to see underwater animals and plants in their natural environment. Kneel down near the edge of the pond or stream and place the plastic-covered end in the water. Don't lean over too far or you could fall in. Now, peer through the other end. What do you see?

If you're feeling really adventurous, go large! Make a bathyscope from a bucket or a garbage can, then place it in the water and see what you can find.

Compare different types of habitats. The critters that live in ponds may be slightly different from the ones that live in moving water, such as streams and rivers. If you live close to the ocean, you could try looking in rock pools. Extra points if you spot any fast-moving fish! Draw what you see and record your findings in your science journal. What is the most exciting animal that you spotted?

THE GREAT SNAIL RACE

Blink and you won't miss it! Find out what surfaces suit snails the best by organizing a snail race.

First, collect five garden snails. Snails like cool, damp, shady environments that are close to food sources. If your garden has a vegetable patch, there are likely to be snails hiding away in the foliage. Try looking under and around plant pots, or around climbing plants that are growing up walls. Handle them carefully and wash your hands once the experiment is done.

Prepare the racetracks. There will be three circular tracks made of wood, cardboard, and soil. Mark out the first two tracks: one on a large piece of wood, and one on a large piece of cardboard. To make the soil racetrack, lay out a circle of soil in a tray. Make sure the circles are all the same size.

Can you feel the excitement mounting? Place the snails in the center of the first wooden racetrack. Time how long it takes each individual to reach the edge of the track. Make a note of these times. Calculate the average time by adding up all the individual times and dividing them by five.

Look at the racetracks. Did the snails leave anything behind? Land snails move using a single muscular foot. They secrete mucus, which helps them to crawl over rough surfaces and prevents their soft bodies from drying out. Can you spot the snail trails?

- Now test the snails on the second track made of cardboard. Place the same snails in the center and time how long it takes each individual to reach the edge. Calculate the average. Repeat the experiment one last time using the third track, which is made of soil. Calculate the average.

- Measure the distance from the starting place to the edge of the track. Now calculate the average speed on each track: speed equals distance divided by time. Ask an adult to help you convert your calculations into a measurement that is in feet (or meters) per second. Which surface were they speediest on?

- See if you can design a racetrack that makes the snails even faster. Would the racetrack be bumpy or smooth, wet or dry? What's the fastest speed your snails can achieve? Record the results in your journal.

WONDERFUL WILDLIFE QUIZ

Once you've tried a few of the experiments in this section, you'll be surprised at how quickly your understanding of the natural world can grow. Test your newfound knowledge with this little quiz.

1. Did you find an activity that you really enjoyed? Name the single, best experiment in this section of the book.

2. Moths can be unexpectedly vibrant and beautiful. Explain two different ways to attract moths to your backyard.

3. Can you describe three different ways to encourage wildlife into your garden?

4. In winter, resources are scarce, so some animals snooze the cold months away. Can you name four different animals that hibernate?

5. Bathe in the beauty of birdsong. Can you recognize five different birdsongs?

6. They're not creepy, but they are crawly. Can you name a garden-dwelling animal that has six legs?

7. Garden birds benefit from additional food. Can you list seven foods that garden birds like to eat?

8. They play a vital part in the garden ecosystem. Can you name a garden-dwelling animal that has eight legs?

9. Ponds are havens for life. Can you draw nine different water-dwelling animals that can be found in ponds?

10. They're a joy to watch and welcome into our outdoor spaces. Can you recognize ten different garden birds?

11. Our gardens are full of life. Can you draw eleven different land-dwelling animals that live in your backyard?

12. Outdoor spaces are full of noise. Can you describe twelve different outdoor sounds you hear every day?

Do you know . . .

- ***How to devise your own experiment and write it up? Scientists always write up their experiments.***

- ***Why birds sing? It's an important part of their life.***

- ***How to build a bug hotel? Welcoming bugs into your garden can be a rewarding experience.***

- ***Where woodbugs like to live? They're not difficult to find. Pick up a plant pot and look underneath.***

- ***How hoverfly larvae breathe underwater? Think of them if you ever go snorkeling.***

- ***Which part of a butterfly's body is used to taste things? Insects are amazing!***

- ***How to collect insects in an umbrella? Just remember to set them free before you use the umbrella on a rainy day.***

- ***What a mouse pawprint looks like? They're very distinctive.***

- ***Where to look for water bears? They live just about everywhere but are so small you've probably never noticed them.***

- ***How to prevent the water in a birdbath from freezing? When the cold comes, the birds will thank you for it.***

- ***How toads catch their food? Can you think of any other animals that catch their food the same way?***

2

SOIL SCIENCE

Soil is really remarkable. The next time you're walking in your garden or local park, stop for a moment and think about the soil that's under your feet. We walk all over it every day and hardly give it a moment's thought. Some people might assume it's dull and boring, but soil is brimming with life and life-giving properties. It's time we learned to appreciate soil and all the things it does for us.

SOIL SCIENCE

Soil is a gloriously grubby mixture of organic and inorganic matter. Organic matter is anything that comes from living things. Fallen leaves, grass clippings, and animal droppings are all examples of organic matter. Inorganic matter is made up of nonliving substances. Pebbles and rocks are inorganic, but soil also contains tiny inorganic particles of things such as sand, silt, and clay.

Soil is amazing for many reasons. It is a life-support machine. It provides plants with the water and nutrients that they need to survive and grow. We need soil to grow the crops that feed us and that feed the livestock we eat.

Soil is a home. It's teeming with living things, such as bacteria, fungi such as the toadstool fly agaric (below), and worms. These

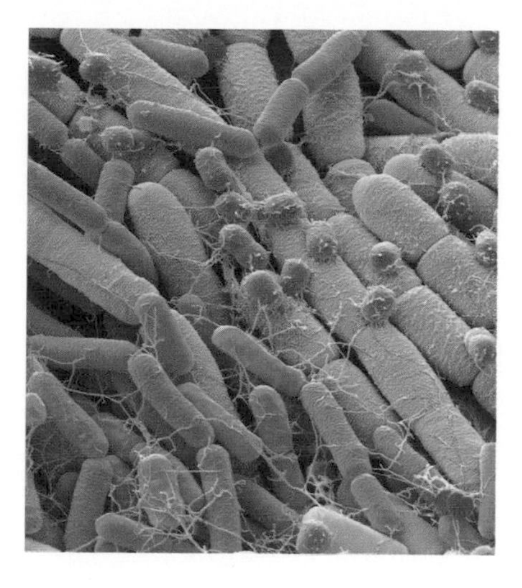

underground marvels work hard to improve the quality of the soil. Bacteria (above), for example, convert nitrogen in the air into soil-based forms of nitrogen that plants can use for growth. Fungi help to break down fallen leaves and dead animals, releasing nutrients into the soil. Earthworms (right) dig tunnels that break up the soil and improve drainage, making it easier for plants to grow.

Soil is a filter. It doesn't only store water. The tiny particles in soil also help to cleanse water of harmful chemicals and pollutants. When plants grow in soil, their roots form tangled networks that help to stabilize the soil and bind it together. This helps to prevent erosion and flooding.

Soil is so much more than dirt. In this chapter, you'll be learning more about the soil in your neighborhood. You'll be testing it to see how acidic or alkaline it is, how much water it retains, and how much air it holds. You'll be learning about its composition and the vital role that soil plays in helping living things to decompose. You'll be making compost, so you can nourish the plants that are growing nearby, and learning about the creepy-crawlies that make soil their home. Most of all, you'll be learning how to appreciate this most grubby and underappreciated marvel—so next time an adult complains about the dirt on your clothes, you can tell them how important it is!

COMPOST JARS

If living things didn't decompose after they died, the world would be chock-full of waste. Fortunately, microorganisms are on hand to break down organic material. Set up some see-through compost jars and time how long it takes for different materials to degrade.

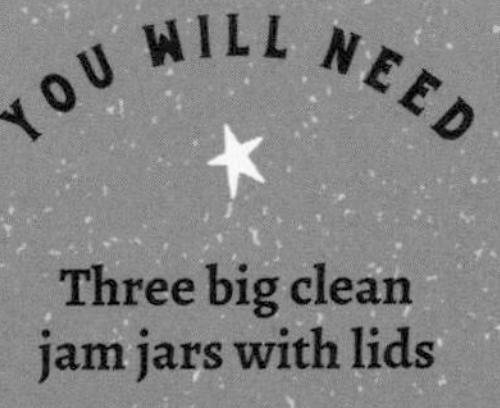

Three big clean jam jars with lids

Soil

Water

Newspaper

Banana peel

Biodegradable plastic bag

Hammer and nail

Thermometer (optional)

- Remove the lids from the jars and make some holes in them by banging the nail into the lid—remove the nail with the claw of the hammer. You might need an adult to help with this task. Make sure you keep your thumb out of the way when banging in the nail!

- Fill the jars halfway with soil from the garden. If the soil is dry, sprinkle it with water, but don't add too much. The soil should not become waterlogged.

- Add the banana peel to one jar, and a crumpled-up piece of newspaper to another. To the third jar, add a scrunched-up biodegradable plastic bag. Many dog poop bags are biodegradable, so these are perfect.

Normal plastic bags can take hundreds of years to decompose, but biodegradable plastic bags don't disappear overnight. They still take months to degrade when they are added to landfill. Think carefully before you dispose of them. You can always put them in the compost bin.

- Fill up the jars with soil, but make sure the fillings are still visible. If you have a mercury thermometer, gently press it into the soil and record the temperature in the center of each jar. Make a note of this. Then screw on the lids and put the jars somewhere warm, such as a windowsill.

- It takes a while for organic material to decompose, so be prepared to wait. Check the jars every week. What differences do you notice? Are the fillings beginning to break down? Unscrew the lid of each jar and take further temperature readings. Note these down. When living things decompose, they release heat, so you should notice an increase in temperature.

- Let the experiment run for two to three months, then empty the jars and see what has happened. Which of the three different fillings has decomposed? Which item degraded first and which took the longest? If an item has yet to decay, put it back in the jar and keep the experiment running. Time how long it takes to decompose.

MAKE IT GROW FASTER

Plants need more than just water and sunlight to grow. They also need nutrients such as nitrogen. Fertilizers contain nitrogen. See how well they work by growing some plants with and without fertilizer.

YOU WILL NEED

- Ten medium plant pots

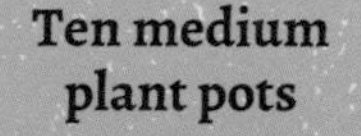

- Potting soil and topsoil
- Packet of sunflower seeds
- Water
- Houseplant fertilizer
- Two large plastic bottles

Add an equal amount of potting soil mixed with some topsoil to each of the plant pots. Fill the pots to the top, then press the soil down with your fingers. Topsoil is just the top layer of soil in the ground.

Choose ten equally sized sunflower seeds from the packet. Add one seed to each pot. Place each seed gently on top of the soil, then cover it up with a thin layer of potting soil. Moisten the soil in each pot by adding an equal amount of water.

Label five of the pots with "fertilizer" and five of the pots with "water." Place all of the pots on a sunny windowsill and leave them to grow.

Label one of the bottles with "fertilizer" and the other with "water." Fill the "water" bottle with water, then prepare the fertilizer. Many common household fertilizers come as concentrated powders or liquids that need to be diluted with water. Carefully follow the instructions on the packet and prepare enough fertilizer to fill the "fertilizer" bottle.

When the soil starts to dry out, the plants need watering. Always add an equal amount of liquid to each seedling. Always water the "fertilizer" plants with fertilizer, and the "water" plants with water. Depending on the temperature, the plants may need watering two to three times per week. Make up more fertilizer as you need it.

Once a week, measure the plants. Make a note of how tall they are and how many leaves they have sprouted. Do you notice anything else? Sunflowers can really shoot up, so as they grow, the plants may need supporting with sticks. If they get really big, they may need moving into a larger pot. If you do this, repot all of the plants at the same time.

Although the air is full of nitrogen, plants can't use it directly. Instead, they rely on soil-dwelling microbes that convert atmospheric nitrogen into other, more plant-friendly forms.

After a few months, analyze the data. Which plants are the tallest? Is there a difference between "fertilizer" and "water" plants? Plot a graph showing how each plant has grown over time.

TEST YOUR SOIL'S pH

pH is a measure of how acidic or alkaline something is. Some plants like to grow in soil that is more acidic, while others like to grow in earth that is more alkaline. Find out what the soil in your garden is like by using this simple test.

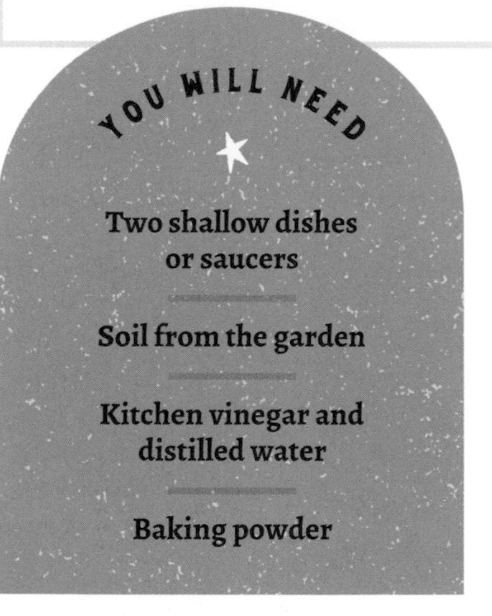

- Grab a handful of soil and rub your hands with it. This will get rid of any products, such as soap or moisturizer that are already on your skin and could affect the experiment.
- Fill the shallow dishes with soil from the garden. The soil should be fairly dry. If it's too wet, put the saucers on a windowsill and let the soil dry out before you continue.
- Sprinkle about five tablespoons of vinegar onto the soil in the first dish and watch what happens next. If the soil fizzes and bubbles, you likely have alkaline soil.
- Sprinkle a tablespoonful of baking powder over the second saucer of soil, then wet the mixture by pouring a little distilled water over it. If it fizzes or bubbles, your soil is very acidic.

If nothing happens when you add vinegar or baking powder, then the soil in your garden is likely to have a neutral pH.

If you saw bubbles in this experiment, they were caused by a chemical reaction. When acidic and alkaline things meet, water and carbon dioxide gas are formed. Bubbles of carbon dioxide rise to the top of the mixture—just like they would in a fizzy drink—and cause the liquid to fizz.

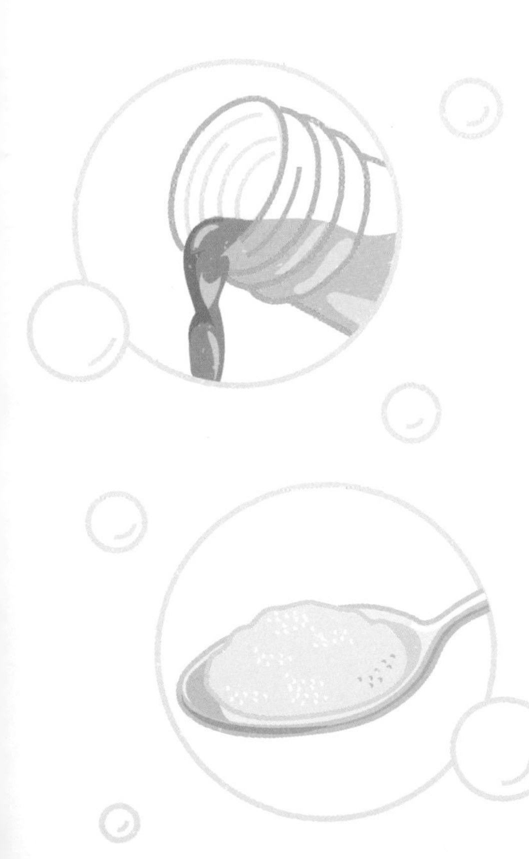

DID YOU KNOW?
pH is measured on a scale of 0 to 14. A pH of 7 is neutral. A pH of less than 7 is acidic, and a pH of more than 7 is alkaline. Most plants like a pH somewhere between 6 and 7.

Look up which plants prefer the soil type that you have in your garden. Could this explain why some plants grow better than others?

If you want to find out the pH of your soil more accurately, try to borrow some pH paper from school. Dip the paper into some watery soil. Wait for the paper to change color, then compare the result to the color chart that comes with the strips. This will give you a numerical value for the soil pH. You could also try making an indicator dye from red cabbage (see pages 152–53).

SOIL TEXTURE EXPERIMENT

Some gardens have light, sandy soil. Others have heavy, clay soil. There are many different types of soil. Find out which type of soil is in your garden using this simple texture test.

Fill the glass jar three-quarters of the way with soil from the garden. Make sure there are no pebbles or large stones in the soil. If there are any big clumps of soil, break them apart with your fingers first and crumble the soil into the jar.

Add a tablespoon of laundry detergent and a tablespoon of salt to the soil. Fill up the jar with water. Stir the mixture using the spoon, then screw the lid on tightly. The ingredients need to be really well mixed, so shake the jar for five minutes.

Put the jar on a windowsill and leave it for a few days. Don't pick it up or disturb the jar during this time. After a couple of days, the soil will settle into different layers.

Soil is made from different types of particles. Sand particles are the heaviest, so they will sink to the bottom. Clay particles are the lightest, so they will settle on the top. Silt is a grainy material. Silt particles are medium-sized, so if there is silt in your soil, it will form a layer in the middle, between the sand and the clay.

Grow plants that are suited to your soil type. For example, lavender plants like sandy soil, while roses like clay soil.

You can estimate how much of each different type of particle is in your soil by measuring the thickness of the different layers. If the sand layer at the bottom is really thick, for example, then you have sandy soil.

Sandy soils are easy to dig in but they dry out easily and are low in nutrients. This means they need regular watering and filling up with fertilizers. Clay soils are heavy and can be difficult to dig in. They are rich in nutrients and retain water well, but can become waterlogged, making it harder for plants to grow. Silt soils are somewhere in the middle.

MAKE A LEAF COMPOST BIN

Compost is made of organic material that has decomposed. It contains lots of nutrients that help plants to grow. Make some leaf compost in a simple compost bin and give your plants a treat.

Four bamboo canes or sturdy canes, about 3 ft (1 m) long

Chicken wire or plastic garden netting

Garden wire

Strong scissors or wire cutters

Gloves

Deciduous leaves

This is an activity for fall when there are lots of leaves on the ground. Decide where you want your compost bin to go. It takes a year or more to make leaf compost, so it is best to choose a spot that's out of the way and won't be needed for anything else.

Mark out how large you would like the compost bin to be. You can make it any size you like, but it doesn't have to be huge. A square shape that is 3 ft × 3 ft (1 m × 1 m) works well.

The bamboo canes are going to form the corners of the square. The canes should be about 3 ft (1 m) tall. Carefully push the canes into the ground so they stand upright without support.

You may need someone to help you with this next step. Unroll the netting and wrap it around the four bamboo canes. If you are using chicken wire, wear gloves for this since it can be sharp. The netting should go all the way round the square once, with a little extra to form an overlap.

Cut some small pieces of garden wire and use them to tie the netting to the bamboo canes. Use at least two ties per cane: one at the top and one at the bottom. Use some extra pieces of wire to tie the two ends of the netting together, where they overlap. Trim any excess using scissors or wire cutters. Your leaf compost bin is ready to go!

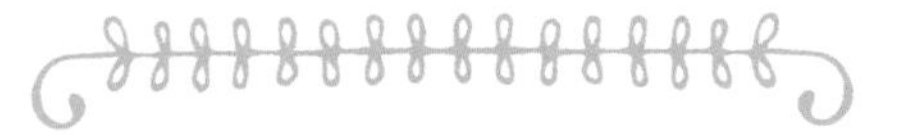

Leaves don't turn to compost on their own. The mixture is broken down by tiny organisms, such as bacteria and fungi.

Collect lots of leaves and pile them up in the compost bin. Use leaves that have already fallen from deciduous trees. Deciduous leaves degrade more quickly than leaves from evergreen trees, so it helps to speed up the composting process.

If the leaves start to dry out, water the compost bin. In a year or two, the leaves will rot away to form a rich, nutritious compost.

SOIL EROSION EXPERIMENT

When soil is washed away, any nutrients or chemicals inside the soil get washed away too. This is a big problem because it can pollute rivers and streams, and harm wildlife. Find out how to stabilize soil and prevent this soil erosion.

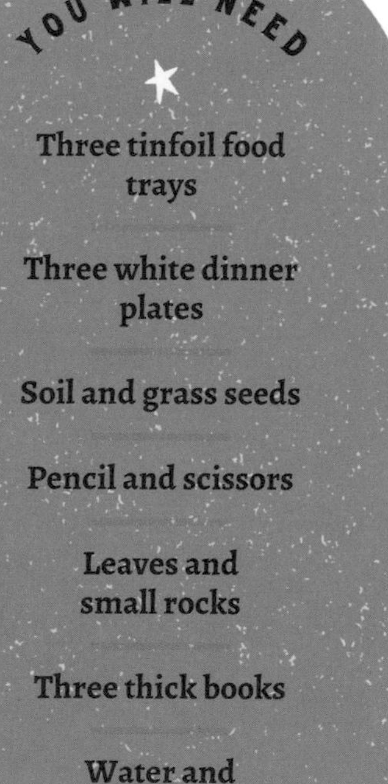

Use a pencil to poke ten drainage holes in the bottom of the three food trays. The holes should be about the same width as a pencil.

Prepare the first food tray. Label it "Grass." Fill it with soil and generously sprinkle the grass seeds on top. Lightly water the seeds so the soil becomes damp. Place the tray on one of the plates and put it on a windowsill. Leave it for a few weeks, and remember to water it regularly. When the grass has grown nice and thick, you can continue with the next stage.

Fill the second and third food trays with soil. Lightly water the soil, so it is roughly as damp as the soil in the "Grass" tray. Label the second tray "Leaves and rocks," then add a layer of leaves and small stones on top of the soil. Label the third tray "Soil only." The third tray is ready to go.

Use scissors to carefully cut off one of the small ends from each tray. This is now the front end. The trays should all now have three sides. Place each of the trays on a dinner plate.

Plant roots form a tangled mesh that helps to keep soil in place. Leaves and stones also help to prevent soil from being washed away.

The trays need to be at an angle, so prop a thick book underneath the back end of each tray. The front end should be pointing downward toward the plate. This simulates a hill.

Now it's time to make it rain. Do this part of the experiment outside. Water the trays, one at a time, using a watering can. Hold the watering can up high so the water is sprinkled all over the tray. Count to ten while you water each tray, then stop.

Look at the water that collects in each of the plates. How much soil is there in each of the plates? Did the grass or the leaves and stones help to prevent erosion?

SOIL FILTRATION EXPERIMENT

Soil acts as a natural filter. It can trap certain pollutants and chemicals, which helps to purify water. "Pollute" water with food coloring, then clean it up with some natural soil filters.

- **Three large plastic bottles**
- **Three glass jars**
- **Stones and scissors**
- **Topsoil and sand**
- **Moss or green leaves**
- **Jug of water and cup**
- **Red and blue food coloring**

Cut the bottoms off the plastic bottles. Turn the bottles upside down and rest them on the open glass jars. Put a tightly packed layer of stones at the bottom of each bottle. The stones need to be bigger than the mouth of the bottle to stop them from falling through.

To the first bottle, add a thick layer of sand (4 in/10 cm). To the second bottle, add a thick layer of topsoil (4 in/10 cm). Topsoil is the top layer of soil that you find in the garden. To the third bottle, add a thick layer of topsoil, followed by an additional layer of moss or green leaves. You can be creative with this last layer. Add in anything from the garden that you think will help to filter the water.

Mix up your food coloring. Add red and blue food coloring to a jug of water to make a deep purple color. This is your pretend pollutant. Pour an equal volume of the pollutant into each of the bottles and watch what happens.

The liquid will trickle through the bottles and some will collect in the jars underneath. This is called the filtrate. If no liquid is coming out, add another cupful of the pollutant to each of the bottles.

Don't be disappointed if the filtrate is cloudy or dirty. This often happens at the start of the experiment. Pour the dirty water out and keep pouring more pollutant into the bottles.

Look at the color of the filtrate in each jar. Which is the most purple? This water still contains "pollutants." It has not been well filtered. Which is the clearest? This is water that has been well filtered.

Soil that is rich in silt or clay acts as a good natural filter. Large particles become trapped in the thick, heavy soil, while water molecules can eventually pass through.

MINI WORMERY

Worms are the unsung heroes of the underground. They break down plant matter, churn up the soil, and enrich the earth with their droppings. Find out about their vital recycling role by making them a temporary home.

YOU WILL NEED

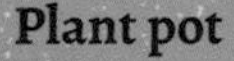

- Plant pot
- Paints and a brush
- Plastic bottle
- Scissors and tape
- Soil and sand
- Grass clippings and a few pebbles (optional)
- Two or three worms
- Newspaper

Brighten up your plant pot by giving it a lick of paint! You could even paint a sign saying "WORMS LIVE HERE." Leave it to dry.

Using the scissors, cut the top and bottom off the plastic bottle to create a see-through plastic tube. This can be tricky so you may need adult help. Be careful! The edges of the tube might be sharp.

DID YOU KNOW?
The Giant Gippsland earthworm of Australia can grow to over 6 ft (2 m) long.

Place the tube in the plant pot so it stands upright. Fill around the outside of the tube with soil to make it more stable. Now fill the inside of the tube with alternating layers of sand and soil. Earthworms like to live in damp earth, so sprinkle each layer with a bit of water. The soil layers should be about the width of your thumb, and the sand layers should be a little thinner.

Leave some space at the top of the tube to add in the grass clippings. Earthworms feed on living matter, so the grass will provide them with food.

With the grass in place, go outside and find some worms. They'll be hiding in the soil. If you can't find them, try soaking the ground with a hose for ten minutes. This will help to bring any earthworms up to the surface. These are living creatures, so handle them with care.

Add two or three earthworms to the top of your wormery, and watch as they bury down into the soil. Worms live in the dark, so wrap the newspaper around the tube and fasten it with a piece of tape. This will make it dark and encourage the worms to visit the edges of the tube where you can see them when you remove the paper cover.

Put the wormery in a cool, dark place and check it each day. Watch how grass disappears and the layers get mixed up. Record in your science journal how long it takes for the grass to disappear. After a few days, return the worms to a safe place in the garden.

HOW MUCH AIR IS IN YOUR SOIL?

It may just look like mud, but soil contains a lot of air. This includes oxygen, which helps plants and animals to breathe, and nitrogen, which helps to nourish plants. Find out the air content of different soil samples using this simple experiment.

Collect three different soil samples. You will need a handful of each. Choose samples from different locations. Perhaps one from a garden, one from a row of hedges, and one from a riverbank. Try to choose samples that look and feel different.

YOU WILL NEED

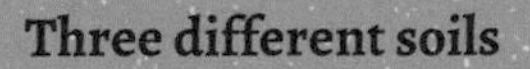

Three different soils

Water and spray bottle

Three glass jars or jugs

Smartphone with camera

If the samples are dry, moisten them using water from the spray bottle. Mold each sample into a ball. The balls should all be roughly the same size and be able to fit inside the glass jars.

Fill the glass jars three-quarters full with water. Drop the first soil sample into the first jar and film what happens next. What do you see? Tiny air bubbles will be released from the soil sample into the water. They float up to the surface and then disappear.

Up to 50 percent of soil is made up of gases, such as oxygen, carbon dioxide, and nitrogen. If the soil is waterlogged or compacted, there is less space for air, and this can make life difficult for soil-dwelling organisms.

It's hard to count the number of bubbles that are released in real time, which is where your video recording comes in. Play back the video—at half speed if you can—and count how many bubbles are released from the sample.

- Now repeat the same process with the second and third samples. Make a table in your journal, detailing the number of bubbles released from each sample. Which sample had the most air? Which sample had the least air? What do you think this means for the animals and plants that live in these different soils?

- Repeat the experiment using soil from the same sites, but this time dig down. Compare soil from the surface with soil that is around 8 in (20 cm) down. Keep digging. Collect a fourth sample that is as far down as you can dig. What do you notice? The soil from the surface should have much more air than the soil from lower down.

HOW MUCH WATER CAN YOUR SOIL HOLD?

It's important for soil to be able to retain water, because plants need water to grow. Different soils hold different amounts of water. Try this experiment to work out how much water the soil from your garden can hold.

YOU WILL NEED

Soil, sand, and clay

Funnel

Three coffee filters

Glass jar

Measuring jug

Kitchen scales

Collect a big handful of soil. This can come from the ground in your garden, or from a plant pot where you grow something. Using the kitchen scales, weigh out 1.8 oz (50 g) of your soil.

Place the funnel into the glass jar and place a coffee filter inside the funnel. Make sure the coffee filter is open. Add the soil to the filter, just as you would add coffee grounds.

Measure out 1.7 fl oz (50 ml) of water in the measuring cup. Slowly pour the water over the soil. Wait for five minutes. What happens? Some of the water will pass through the soil and drip through the funnel into the jar underneath. Pour the water into the empty measuring cup. How much did you manage to collect?

- Now repeat the experiment with a fresh soil sample, but this time find out what happens when you add some sand to the soil. Sand is part of the natural composition of soil, but some soils are sandier than others. Did you collect more or less water this time?

- Repeat the experiment one last time, with another fresh soil sample. This time crumble some dried clay into the soil. Clay is also found in many soils. How much water did you collect during this final part of the experiment?

- Make a note of the results in your journal. Why do you think the experiments gave different results?

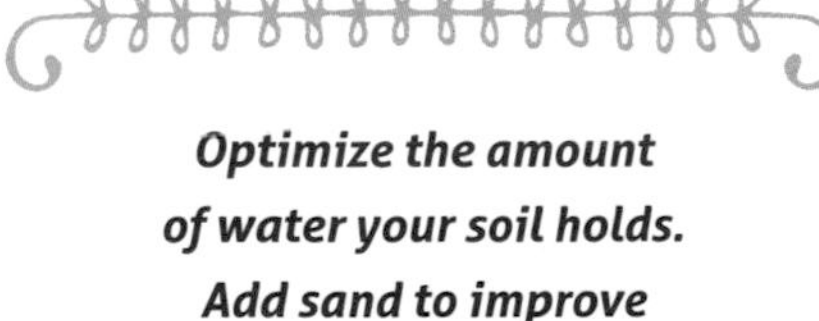

Optimize the amount of water your soil holds. Add sand to improve drainage. Add compost or clay to improve water retention.

Sand particles are larger than clay particles, so there are more spaces, or "pores," for the water to filter through. So, if you want to improve the drainage in your garden, add some sand. Clay particles are smaller than sand particles. They sit snugly next to one another, so there aren't many pores at all. This means that clay soils retain more water than sandy soils. This can be handy if you live somewhere hot, where it doesn't rain much.

FANTASTIC FACTS

It's gloriously grubby and downright dirty, but soil is a haven for life and a crucial part of many different ecosystems. It might seem brown and boring, but soil is also packed full of interesting facts. Here are a few that will impress any soil skeptics that you might know.

★ One-quarter of all the world's known species live in soil. That's millions of different species, but only 1 percent of microscopic soil-dwelling species have been identified. If you look carefully, you might even discover a new species!

★ One teaspoon of soil contains thousands of different species of living things, but most of them are so small, they are invisible to the naked eye. This includes microscopic creatures such as bacteria and some fungi.

★ Soil takes many hundreds of years to form. Soil is formed when rocks get broken down into tiny microscopic pieces. This is called weathering. Rocks are responsible for the minerals that are found inside soil.

★ Healthy soils aren't always brown. Depending on the minerals and organic matter present, they can be red, yellow, black, white, green, or gray. Brown soil contains lots of organic matter, while greenish soil contains a mineral called glauconite.

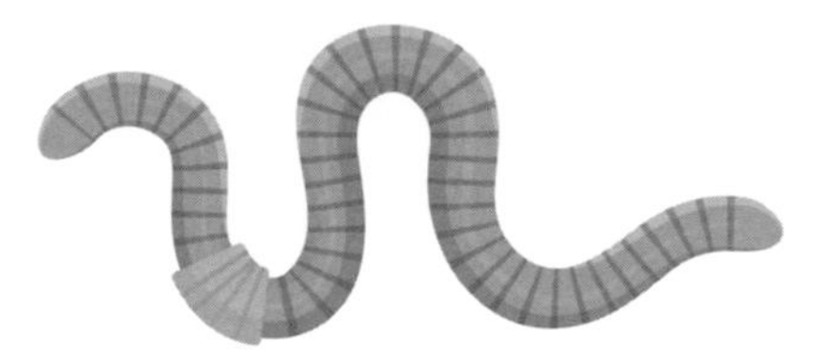

★ British scientist Charles Darwin loved worms. He called them "nature's ploughs" because of their ability to mix soil and organic matter, and he tested their ability to hear by playing them his bassoon.

★ Worms are among the most recognizable soil dwellers. There are more than 5,000 different species of worm. The Fried Egg Worm (above) from the Philippines looks like it has tiny fried eggs scattered along its body.

★ We all know that trees are important because they help to lock away atmospheric carbon dioxide, but did you know that soil stores more carbon than all the world's forests? When living things die and rot away, the carbon they contain is transferred to the soil.

★ Without soil, we'd be very hungry. Ninety-five percent of the food we eat comes from the soil. You can add eggshells, tea bags, and coffee grounds to your compost pile. They contain lots of molecules that help plants to grow.

★ When forests are cut down to make way for big industrial farms, the crops use up the nutrients and the soil becomes less fertile. This puts soil in danger. Every minute around the world we lose the equivalent of thirty soccer fields of soil. We need healthy soil to feed the world.

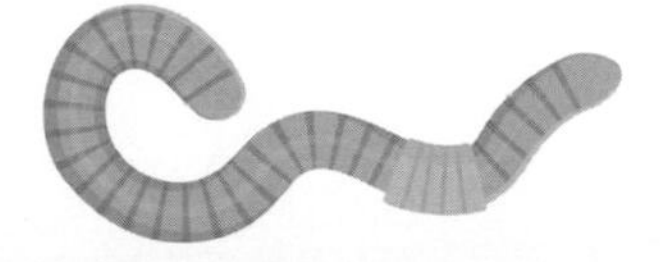

3

FASCINATING FLORA

If you were to weigh all of the living things on our planet—including animals, plants, fungi, bacteria, and viruses—you would find that you need a very big pair of scales. In addition, you would also find that plants make up more than 80 percent of the total amount. We live on a planet covered in plants. They are vital to our survival, so it's time we learned to love them.

FASCINATING FLORA

Plants are phenomenally successful. They grow just about everywhere: in the heat of the desert (see beavertail cactus, opposite), in the icy wastelands of the Arctic (see cotton-grass, opposite below), and in the inhospitable cracks and crevices of buildings and streets.

The smallest plants are made of a single cell, while the tallest plants—evergreen redwood trees—tower more than 330 ft (100 m) above the ground (see below). Plants give us glorious displays of color and inviting places to explore. They provide us with food, medicines, and building materials, but more than that, they give us the very oxygen that we breathe.

Powered by sunshine, plants absorb carbon dioxide from the air, combine it with water, and convert it into sugars and oxygen. The sugars are used by the plants to

you work out the age and height of your favorite tree, and understand why it is that leaves change color in the fall. You'll be growing plants from food scraps and edible seeds, and making leaves disintegrate so you can see their skeletons. You'll be creating vegetable animals, making flowers change color, and growing your name in garden cress. One minute you'll be lovingly growing seedlings in a greenhouse, and the next you'll be smashing up plants with a hammer. It's time to enjoy the wonderful world of plants!

help them grow, while the oxygen is released into the atmosphere. This process is called photosynthesis and it's the reason some people call trees the "lungs of the earth." They help the planet to breathe.

As humans pump carbon dioxide into the atmosphere and rain forests are felled to make way for food crops, we need trees more than ever before. They help to mop up carbon dioxide, and so can help us to control climate change.

It's time we appreciated how important plants are. In this section, you'll explore how plants grow and the different things that they need to keep them strong and healthy. There are experiments to help

PHOTOTROPISM BOX

Plants need light to help them grow, so they always grow toward it. This is called phototropism. Perform an experiment to demonstrate phototropism using a cardboard box, a potato, and some sunshine.

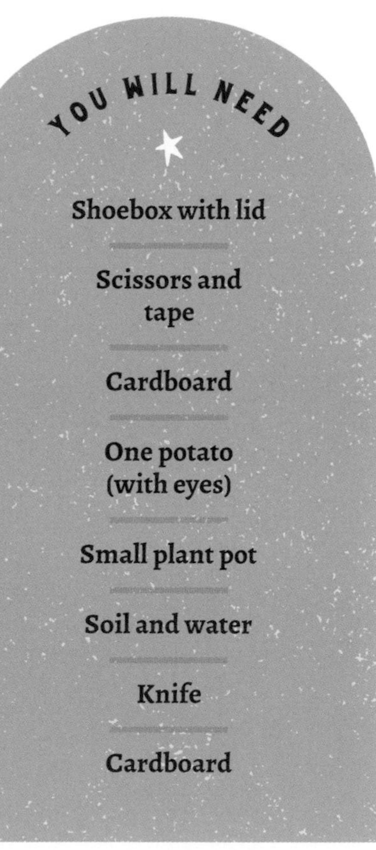

Stand the shoebox on its shortest side so it stands upright, long and tall. Using scissors, make a hole in the middle of the "roof" of the shoebox. The hole should be about 1 in × 1 in (3 cm × 3 cm).

Make two cardboard shelves to fit inside the shoebox. Mark out the shelves on the cardboard by drawing around the base of the shoebox when it is standing upright. Cut them out. The shelves now need to be trimmed. Cut 1½ in (4 cm) off the end of each rectangle.

Fit the shelves inside the shoebox using tape. The two shelves should split the box into three equal parts. The bottom shelf should be fixed to one wall of the box, and the top shelf should be fixed to the opposite wall.

The shelves will prevent the light from falling directly onto the potato plants and create obstacles that the shoot needs to grow around.

★ Prepare your plant. You need an old potato that has started to sprout eyes. The eyes are the beginnings of new shoots that are starting to grow. Fill the plant pot with soil. If the soil is dry, water it a little, but take care the soil doesn't become waterlogged. Cut the potato in half and press the freshly cut side down onto the soil.

★ Place the plant pot into the bottom of the shoebox. Close the shoebox and secure the lid by putting tape around it. Put the shoebox on a warm windowsill and leave it for a couple of weeks. Once a week, open the box up and check to see if the plant needs watering. Then seal it back up again.

DID YOU KNOW?
Photo means light. Plants contain special hormones that help them to grow toward the light.

★ After three weeks, you should see a green shoot poking out of the hole at the top of the box. The experiment is now complete. Draw a diagram to show how the plant grew. Did the shoot grow around the shelves to get to the light?

HAPA ZOME

Hapa Zome is the Japanese art of beating leaves and flowers with a hammer in order to make natural prints on fabric. Give it a go! It's easy and effective, and the results are beautiful.

Go outside and choose some leaves and flowers. They should be as varied as possible, in different shapes and sizes. Choose a variety of colors. White flowers won't make a mark on white fabric, so pick blooms with bright or dark-colored petals. The leaves and flowers should be freshly picked. Ones that are full of moisture work the best, so make sure you collect fresh, healthy specimens.

Prepare your fabric. Plain white fabric works, but you could go for any light-colored cloth. Thin cotton fabrics work well, so you could cut up an old pillowcase or bedsheet (but make sure you check with an adult first). Cut a piece of fabric that is the size of the image you want to make. Now cut a second identical piece.

Lay the first piece of fabric on a wooden cutting board or other flat surface. Position the leaves and flowers on the fabric, arranging them into the design that you would like to see printed. Now lay the second piece of fabric over the top.

Here's where the fun starts. Gently tap the fabric with the hammer. If you don't have a hammer, or aren't allowed to use one, a rolling pin or smooth rock works just as well. Be gentle. Don't batter the fabric. Tap at it lightly. What do you see? The pattern of the plants should begin to appear.

Once you have flattened all of the plants, peel away the top layer of fabric. Remove the leaves and flowers. The dye from the plants will have transferred onto the two pieces of cotton. Hang them up to dry, then enjoy!

What will you do with your Hapa Zome prints? If you want to make decorative flags, you can cut them into triangles and thread them onto a piece of string. Alternatively, the prints make great wall hangings, and if you stick them onto cardboard, they make stunning greeting cards.

MAKE A PLANT PRESS

You can keep a record of the plants that grow in your garden by pressing and drying them. Pressed flowers and plants are also really great for using in arts and crafts projects. It's easy to make a simple press out of recycled materials.

Newspaper (small sheets are easier)

Corrugated cardboard

Two flat pieces of wood, of equal size

Two belts, luggage straps or bungee cords, or string

Freshly picked garden plants

- Fold the newspaper sheets down the middle, where the fold normally is. These newspaper sheets will be your "blotters," soaking up the moisture from the plants. Assemble them into sets of three, so that you can open and close each set together, like a mini newspaper.

- Cut rectangles of cardboard so that they're the same size as your folded newspaper sheets. The tunnels in the corrugated cardboard will allow air to flow through your press.

- Start by putting a sheet of cardboard on top of the wood base, then a set of three

newspaper sheets. Open up the sheets and place a plant in the center, then close them so the plant has three sheets of paper on either side.

- Add another piece of cardboard to the stack, and then another set of newspaper sheets, and another plant. Keep going until you've put all your plants into the stack.

- Finish the stack with a last layer of cardboard and then place the second piece of wood on top.

- Now you need to put the straps around the press to hold it tightly closed. You can tie the press together with string, like a parcel, if you don't have straps.

Leave the plants in the press for twenty-four hours, and then you can check on them if you want to. You'll find that they take a few days to dry completely, and then you can reuse your press to flatten more plants—remember to use fresh newspaper.

Botanists press specimens of plants they find on their travels around the world, and these are kept in special libraries called herbaria. A herbarium can tell us a lot about where different plants grow.

FLOWER DISSECTION

Some flowers are simple, others are complex. More complicated flowers contain rows and rows of petals, all tightly packed around a central core. Dissect a flower to create a work of art, and learn how the different parts of a flower fit together.

- Get your hands on some fresh flowers. These can be cut flowers from the supermarket, or wildflowers from the garden. Choose a variety of different flowers, but be sure to include some with large bright petals.

DID YOU KNOW?
The world's biggest flower is the* Rafflesia. *It comes from the rain forest of Indonesia where it can grow up to 39 in (1 m) across. Imagine dissecting that!

- Flowers have lots of different parts. The sepals are the small green leaves on the outside of a blossom. They help to protect the developing flower buds. Draw a stalk on your piece of paper, then glue the sepals around the top of the stalk in a semicircle.

- Petals are actually modified leaves. Pull off the petals and stick them in a semicircle around the sepals. If you have lots of petals, you'll need to make extra rows of semicircles that radiate outward.

- Is there anything of the flower left? The inside contains the flower's reproductive parts. Most flowers have male parts called stamens and female parts called carpels. Stamens

YOU WILL NEED

Flowers

Large pieces of paper

Glue

Patience!

are long filamentous structures that have blobs of pollen at the end. The carpel is the central structure that contains the ovary. Carefully pull off the stamens and carpel, and glue them onto the paper. This is the final, outer row.

- Repeat the experiment using different types of flower. Do you notice a difference between plants that are insect-pollinated and plants that are wind-pollinated?

- Insect-pollinated plants have brightly colored petals, which help to attract pollinators such as butterflies and bees. The petals of wind-pollinated plants are often dull green or brown because they don't need to attract insects.

- The stamens of insect-pollinated plants are often stiff and firmly attached to the inside of the flower, so the insects can brush against them. The stamens of wind-pollinated plants often dangle down outside the flower, so they can release their pollen easily when there is a breeze.

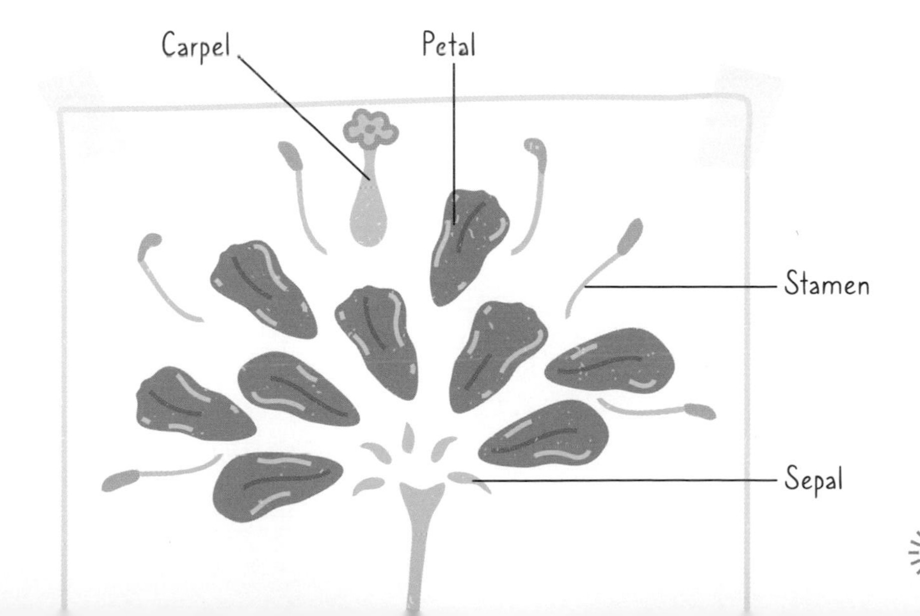

MAKE A VEGETABLE ANIMAL

Let your imagination run wild with this test of creative thinking. Create a vegetable animal using only local fruit and vegetables that are in season.

These days, it's all too easy to buy seasonal food items, such as raspberries and lettuces, all year round. Often the items are grown far away and then transported vast distances to our supermarkets. This requires a lot of energy, making fruit and vegetables that are produced in this way less environmentally friendly than seasonal, locally grown produce.

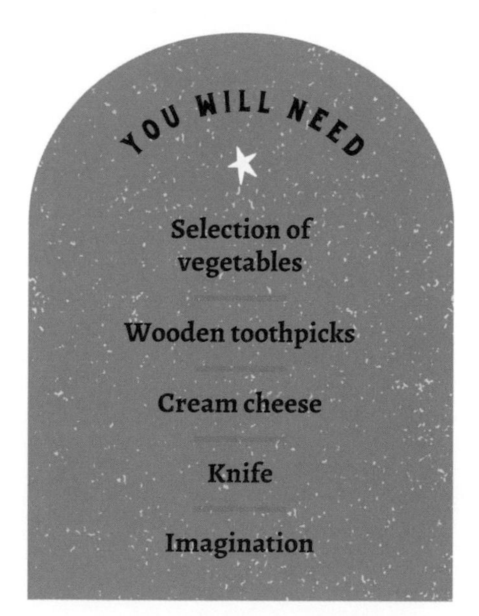

Find out what fruits and vegetables are available locally. If you grow fruit and vegetables in your garden, ask a grown-up if you can use them for this project. If you don't, head to a farmer's market where local, seasonal products will be on sale. To keep costs down, ask vendors if you can have produce that is past its best. They may give it to you for free.

If you want your creation to last, avoid squishy foods such as grapes and tomatoes. Hard vegetables such as potatoes and carrots take much longer to degrade.

Build your animal. Use wooden toothpicks to attach different body parts together, and cream cheese to "glue" on any small features, such as eyes or teeth. Sculpt different body parts and cut them into shapes. Be creative. It's great to build cats and dogs, but can you build something more unusual, such as a stegosaurus or an octopus?

How easy was it to find fruit and vegetables to use? Take a photo of your creation, then eat your animal! Any fruit can be guzzled directly, while any vegetables can be turned into soup. Ask an adult to help you with this. Don't let your vegetable animal rot away on a bedroom shelf. What a waste that would be!

Try the same task again three months later when the season has changed. What can you make this time? As different ingredients become available, you may find that your inspiration changes.

Make this an engineering challenge by trying to build the tallest free-standing animal that you can.

ROOT-GROWTH VIEWER

Plant roots always grow in the direction of gravity. This is called gravitropism (or geotropism). Watch gravitropism in action by building a root-growth viewer from an old CD case.

YOU WILL NEED

- Old CD case
- Bowl
- Soil and water
- Beans
- Tape
- Eyedropper
- Permanent marker

Find an unwanted CD case that is made entirely of transparent plastic. Open up the CD case and remove the plastic tray that usually holds the CD. This can be recycled.

Place a large handful of soil in a bowl. Crumble the soil with your fingers to get rid of any big lumps. Mix the soil with a little water so that it is moist but not too soggy.

Fill the CD case halfway with the damp soil, leaving the half closest to the hinge empty. This will give your bean room to grow.

Place the bean in the middle of the soil, then sprinkle it with water. Green beans and fava (broad) beans work well. Close the CD case and stand it upright so the hinge is at the top. Tape around the sides of the case, but do not tape up the gap in the hinge. This will provide the plant with air and an escape route for the shoot when it becomes too big for the case.

- Prop up the CD case on a sunny windowsill. Water the bean daily using an eyedropper or syringe by adding a few drops of water through the gap in the hinge.

- Look at your seed every day. As the seed starts to grow, mark its growth on the CD case using a permanent marker. Label the seed coat, roots, shoots, and leaves. What happens to the seed coat as the plant gets bigger? It should begin to shrivel as all the food reserves that are inside become used up by the growing plant.

DID YOU KNOW?
In space, there is very little gravity, but plants can still grow. Scientists have grown cress onboard the International Space Station.

- The roots will be growing down. When the roots are around 2 in (5 cm) long, turn the CD case 90 degrees and prop it back up on the windowsill. The roots will now be pointing in the wrong direction, but watch what happens to the roots over the next few days. They will reorient and start growing in the direction of gravity. Draw a picture of your results in your science journal.

SENSORY GARDEN TEST

Gardens can be so much more than just grass and bushes. With a little thought, it's possible to create an area that stimulates all of the senses. Test your senses with a homegrown sensory garden.

Find a little bit of garden where you are allowed to grow things. If you don't have a backyard, you can grow a sensory garden in a large pot or window box.

This activity is all about choosing plants that tickle the senses in different ways. You are looking for plants that will give you something to touch, taste, hear, smell, and see. To keep costs down, choose plants that can be grown easily from seed, or ask friends for cuttings of plants they already have.

Texture is easy to find. Leaves vary a lot, from rough to smooth, and furry to spiky. Plant something you will enjoy touching, such as the lamb's ear, shown opposite. These are small silvery plants with soft velvety leaves that feel just like lambs' ears.

YOU WILL NEED

A little bit of garden

Seeds and plants

Digging tool

Water

Sunshine!

Herbs are tasty plants. Sow basil or chives from seed. They grow quickly and are a welcome addition to many recipes. Alternatively, you could grow vegetables to tickle your taste buds. Lettuces, peas, and carrots are all fast and easy to grow.

It's not easy to find plants that make a noise, but they do exist. Tall grasses make a whispering noise when the wind blows through them. Corn and bamboo also grow big and tall. Listen carefully and you will hear them rustle on a windy day.

Many plants are strongly scented. Some smells help to attract pollinators, while others act to deter predators. Try planting lavender (which smells like lavender), curry plants (which smell like curry), and the wonderful chocolate cosmos flower (which smells like everyone's favorite treat).

After planting for texture, taste, sound, and scent, add the plant that you most like to look at. Beauty is in the eye of the beholder, but who doesn't like a sunflower? These jolly giants are easy to grow, stunning to look at, and attract insects and birds to the garden.

DID YOU KNOW?
The corpse flower, which grows in Sumatra, is one of the world's stinkiest flowers. It smells like rotting flesh, but thankfully, it only blooms for around twenty-four hours, once every four to six years. Don't plant one of these in your garden!

Record all of the plants that you grow and the senses that you use to enjoy them. Can you tell the plants apart with your eyes shut?

COLOR-CHANGING FLOWERS

Plants absorb water from the soil, then transport it from their roots all the way to their flowers, leaves, and shoots. This is called transpiration, and you can demonstrate it by keeping a cut white flower in food coloring. Watch what happens to the petals.

YOU WILL NEED

- A selection of small vases or glass jars
- Water
- Food coloring
- Cut white flowers
- Sharp knife

Fill the vases with fresh water. If your cut flowers come with plant food, mix up a jug of water with plant food and then fill the vases with the mixture.

Add a few drops of food coloring to each vase so the water changes color. Gel food colorings tend to work better than natural food colorings, and darker colors, such as red and green, tend to work better than lighter colors, such as yellow and orange. Make the water in each vase a different color.

Experiment with different flowers. You could even try to make a lettuce leaf or a celery stick change color. It's also fun to split a flower stem down the middle, then put the two halves in two vases with different colored water. What do you think will happen?

Trim about 1/2 in (1 cm) off the stem of each flower before putting a single flower into each vase. You can use any cut white flower, but roses, chrysanthemums, and carnations (shown here) all work well.

Place the vases on a table or shelf and watch what happens. Cut flowers have no roots, so the water is instead drawn into the plants' stems. The stems are full of tiny tubes that are made from a plant tissue called xylem. The water flows through the tubes all the way to the flowers.

You can see when the colored water reaches the flower because the flower starts to change color. This can happen quite quickly. Chrysanthemums start to change color in just half an hour. They look stripy. This is the food coloring inside the xylem tubes in the flower.

Let the experiment run for a week or more. Change the water every few days. Every time you do this, cut another 1/2 in (1 cm) off the stem of each flower. This helps to keep the liquid flowing through the plant. Take a photo of the results. How bright are your flowers?

GROW YOUR NAME

Garden cress is simple and fun to grow. The tiny seeds contain everything that the plants need to germinate. Grow your name in cress, then eat the results!

You will need as many paper towels as there are letters in your name. If your name is Helen, for example, you will need five paper towels. If your name is Joe, you will need three. Spread out the paper towels on the tray. If they don't fit, fold them in half.

Write your name on the towels in big letters. Write one letter per towel, so your name is spread across the tray. Spray the towels with water so they are damp. Don't overwater the towels or they will fall apart.

Carefully sprinkle the garden cress seeds over the letters of your name. Press them lightly into the paper towels using your fingers. Take care not to spill any seeds in the area around your name. Place the tray in a sunny indoor spot, such as on a windowsill.

Water the seeds regularly using the spray bottle. Make sure the paper towels don't dry out. If it's warm, you may need to water the seeds twice a day.

The seeds should sprout quickly, within a day or two. About one week later, the plants should be full-grown. Place the tray on the ground and take a photo of your handiwork from above. How does your name look?

DID YOU KNOW?
Garden cress seeds can be eaten too. They're extra delicious when baked or toasted.

When you're done, use scissors to trim the cress and eat the green clippings. Garden cress is packed full of vitamin C and minerals, such as potassium and iron. It has a mild, peppery taste and makes a great addition to sandwiches, salads, and soups.

If you let the garden cress grow until it's around 2½ in (7 cm) tall or taller, then it should grow back when you cut it. This means that your name will reappear in all of its green glory. You'll be able to do this a couple of times before the plant goes straggly and starts to make seeds rather than leaves.

GROW PLANTS FROM FOOD SCRAPS

Plants have such a strong urge to grow, that some plants we eat will keep on growing from the parts that we throw away. Some of them will even grow into completely new plants you can eat! See which ones you can grow on the windowsill.

YOU WILL NEED

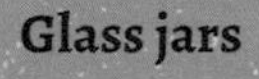

Glass jars

Plates or saucers

Fresh water

Food scraps: carrot tops, onion bases, celery or lettuce bases, and leafy herb leaves

For root vegetables such as carrots and beets, you want to keep the tops of the root (where the leaves attach to the vegetables) and a little bit of the root itself. To get them to grow, put them onto a saucer of water. Check every day and add more water to the saucer as it dries out. In a few days you should see fresh green leaves growing. You can eat them.

Beet leaves have a mild flavor and make a nice salad. Carrot leaves can be a bit bitter, so you may not like them.

For onions, you want the bases of the bulbs (where the roots stick out). You can use the bottoms of spring (green) onions, or bulb onions; it doesn't matter. Pop them into the bottom of a glass jar with some water. Check on them every day and replace the old water with fresh water. Very soon you will see new roots growing from the onion bases, and you may start to see fresh leaves as well. The leaves are edible, just like chives.

Plants are very different from animals, because they can often regrow after being cut in half. They need air, water and sunlight to regrow, but they will eventually run out of nutrients unless they're planted in soil.

You can regrow the leafy greens that grow from a solid section (called a "heart"), such as cabbage and lettuce. We often throw that part away because it's tough, but if you put it onto a saucer of water and keep checking on it every day, you should find that it starts to grow new roots and fresh, edible leaves.

You can also try growing the leaves themselves. Leafy greens—and leafy herbs such as parsley, cilantro, and watercress—will often grow roots in a jar of water. If you want to, you can then pot them up into soil, and they will continue to grow into larger plants.

SEED BOMBS

Welcome wildlife into your garden by creating new habitats full of wildflowers and grasses. Make bombs filled with wild seeds, then launch them at a bare patch of earth.

This is an activity for spring or early summer, when the ground is soft and there is plenty of rainfall. Collect some wildflower seeds. These can either be seeds that you harvest from wildflowers growing in the garden or packets of seeds bought from the store.

Think about the place where you want the seeds to grow. Is it shady or sunny? Different plants like to grow in different places, so choose a combination of seeds that will grow well in the site you have selected.

Prepare the bombs. Add one handful of seeds, three handfuls of clay, and five handfuls of compost to the mixing bowl. Powdered clay can be bought from craft stores, but some gardens have soil that already contains a lot of clay. You can tell if your soil is rich in clay by testing its texture (see pages 74–5) or just by feeling it. Clay soil is sticky. It doesn't crumble easily and can be very difficult to dig! If you have clay soil in your garden, use a slightly different recipe: one handful of seeds, five handfuls of clay soil, and two handfuls of compost.

It's time to get your hands dirty. Mix up all of the ingredients in the bowl, then add some water, a little bit at a time, until the mixture becomes sticky. Break off handfuls of the mixture and mold it into balls using your hands. Place the balls somewhere warm, such as on a windowsill, and leave them to dry overnight.

Now for the fun bit! "Plant" the seeds by lobbing the bombs at bare patches of earth in your backyard. Ideally, the seed bombs should be detonated the day before rain is forecast. A little water will help to give the seeds a head start.

DID YOU KNOW?
Some birds drop seed bombs of their own when they eat seeds and then poop them out!

See how high or far you can throw your seed bombs before they explode, then sit back and watch the plants grow. Make a note of your successes in your science journal.

SPEEDY SEEDS

Seeds germinate too slowly to watch in real time, but when the process is recorded and sped up, the seeds seem to come to life. Make a time-lapse film of seeds germinating and watch as the shoots reach for the sky.

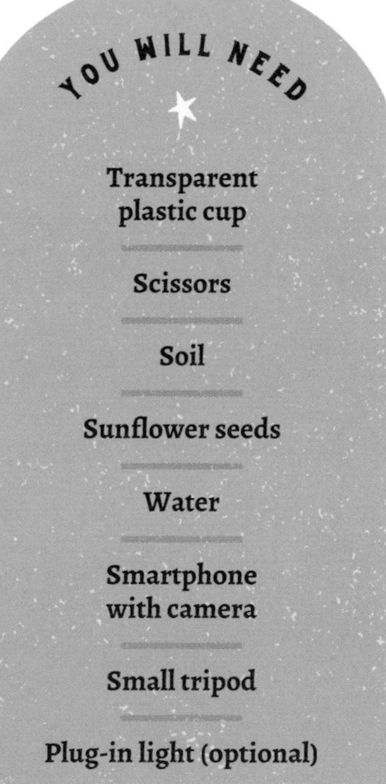

Create a couple of drainage holes in the bottom of the clear plastic cup by pushing the tip of the scissors through the base. Fill the cup halfway with soil. Carefully place a couple of sunflower seeds at the edges of the cup, and cover them lightly with ½ in (1 cm) of soil. Make sure the seeds are still visible through the side of the cup. Water the soil lightly.

Get ready to make your time-lapse video. All videos are made up of fast-changing pictures, or "frames." In order for our brains to perceive these pictures as moving, they need to be played back at a certain speed. Most videos play at least twenty-four frames per second.

Time-lapse videos still play twenty-four frames per second, but all of the frames are recorded separately with a gap between them. So, a filmmaker may take one picture, wait for five minutes, take another picture, wait for five minutes, take another picture, and so on.

This can be very time-consuming! Fortunately, most smartphones and digital cameras have a built-in time-lapse function. This means you can set the camera up and then leave it to take the pictures for you.

Set up your experiment somewhere light and bright, such as a windowsill. It's really important that the camera doesn't move or wobble during the filming, so it needs to be mounted on a tripod.

If you're inspired by this experiment, try taking a time-lapse film of a flower opening.

Put the cup and the tripod on the windowsill, then adjust the height of the tripod so the camera is level with the seed.

The sunflower seed should be in the middle of the picture with plenty of space all around it. That way, when the roots and shoots emerge, you'll be able to record them growing. Make sure the shot is in focus, then start recording. Don't forget to water the sunflower as needed.

GROW A MINI GARDEN

Many people don't have a big outdoor garden of their own, but it doesn't matter. It can be fun to grow your own mini garden in a seed tray. See what mini things you can manage to grow.

YOU WILL NEED

Paper and pens

Seed tray

Compost

Mini plants

Moss (optional)

Sticks, pebbles, and craft items

When professional gardeners are making a new garden, they don't just turn up and start digging. They begin by designing their garden with a pen and paper. Think about what sort of miniature garden you'd like to create. It could be a regular family backyard with grass and a patio, or it could be something different. Perhaps you could make a wildlife garden or a prehistoric garden full of dinosaurs. Plan your design on paper and think about how you will create the different features that you want to include. Remember it all has to fit inside a seed tray.

When you are ready to begin, fill the seed tray with compost. Make sure the compost is level and water it lightly. Following the plan, add in your features. Gravel can be used to make paths, lids can be used to make ponds, and twigs can be used to make fences or wigwams.

If you're feeling adventurous, you could include a larger structure such as a shed or a stream.

The plants also need to be mini. Grow mini trees from seeds or take cuttings from living plants and push them into the soil. Slice the top off a carrot or a parsnip and place it in water (see pages 112–13). In a few days it will have sprouted into a "tree" that you can include in your garden.

To create a lawn, sow grass seeds directly onto the compost. If you want a faster alternative, collect moss and press it into the soil. Hey presto, instant grass! Add in plants with tiny flowers such as daisies or forget-me-nots.

If you want to make it extra realistic, make mini birds and tiny insects from modeling clay and then put them in. Perhaps there could be chickens in the garden or fish in the pond? What else can you include in your miniature garden?

The mini garden is alive, so remember to water it regularly and give it a trim when the plants get too big.

SOW KITCHEN CUPBOARD SEEDS

Our kitchen cupboards are full of dried seeds that we use for cooking and flavoring dishes. It's easy to forget that these ingredients are plants that can still grow with a little help. Rummage through your kitchen cupboards and sow some cupboard seeds. Which ones grow the best?

- Take a look through your kitchen cupboards and find some dried seeds. Canned items that have been precooked won't grow, but dried seeds, which usually come in plastic bags or cardboard packaging, should work well.

- See how many different types of seeds you can find. Sometimes these seeds are sold as single ingredients. You may, for example, find packets of dried chickpeas, lentils, mung beans, buckwheat, and barley. Sometimes these items

Try to harvest and eat the crops that you grow. Peas are delicious, but so are pea shoots, which are ready to eat in just a few weeks.

YOU WILL NEED

- **Kitchen cupboard seeds**
- **Paper towels**
- **Tray**
- **Water**
- **Plant pots**
- **Compost**

come all mixed up. Some retailers sell mixed packets of seeds that people use to make soups.

Check the packaging to make sure the dried seeds are not split. Split lentils, for example, are lentils that have literally been split in half. These will not germinate.

The seeds have been dried out to help preserve them. Many store-bought seeds will germinate, but they need to be rehydrated first. Lay out some paper towels on a tray. Moisten the towels with water. Place five or six of each of the different seeds onto the towels, then cover the seeds with a layer of paper towels. Moisten the top layer of paper towels with water. Place the tray somewhere warm.

The different seeds will rehydrate and germinate at different rates, so check the seeds every couple of days. They need to be kept moist while they are germinating, so keep sprinkling the towels with water.

When the shoots start to appear, transfer the seeds into plant pots. Fill the pots with compost and gently place the seedlings on top. Sprinkle a little more compost on top, but make sure that the shoots are poking through. Water the plants regularly and keep a record of which seeds germinate successfully. In time, you will have a windowsill full of crops.

LEAF RUBBING

Leaves come in all different shapes and sizes. See how many different leaves you can find, and make your own indoor tree out of the leaf rubbings you make.

Collect as many different leaves as you can find. What trees are there in your backyard? Are there different trees in the local park? Gather a variety of leaves in different shapes and sizes. This is an activity for spring, summer, or early fall, when there are leaves on the trees. Leaves lying on the ground can be used to make rubbings as long as they are still fresh and firm.

Place one of the leaves onto the book. (The book is just there to provide a hard, smooth background surface.) Make sure the smooth side of the leaf is facing down, and the veins on the underside of the leaf are facing up.

Cover the leaf with a piece of white paper. Weigh the paper down by placing a couple of heavy pebbles at the edges. This will help to stop the paper from moving while you are doing your rubbing.

DID YOU KNOW?
The world's longest leaves belong to a palm tree called* Raphia regalis*. Each leaf can be more than 10 ft (3 m) wide and 78 ft (24 m) long, but the structure is subdivided into around 180 separate leaflets.

YOU WILL NEED

Lots of leaves!

A big book (any will do)

Paper

A couple of heavy pebbles

Colored crayons or soft pencils

Adhesive putty

Grab your crayons or pencils and lightly rub around the outline of the leaf. You should see the shape of the leaf on the paper. Now rub over the inside of the leaf, taking care to highlight the veins and patterns that emerge.

Make lots of leaf rubbings, using different colors and different leaves. Cut the leaves out and put them to one side. Now it's time to make the tree trunk.

Go outside and feel some tree trunks. Find a tree with a textured trunk. Carefully hold a piece of paper over the trunk and make a rubbing of it. You'll need to repeat this several times in order to make a long tree trunk.

Assemble your tree. Find a bare wall that needs livening up, but check with an adult first. Start by sticking on the trunk, then arrange the leaves to complete the tree.

HOW TALL IS YOUR TREE?

You don't need a tall ladder to measure the height of a tall tree. It may sound odd, but a simple way to do this involves bending over and sticking your head between your legs. Give it a try to find out how tall the trees in your garden are.

Find the tree that you want to measure. Stand with your back to the tree, then walk forward a couple of paces. Keeping your legs straight, bend down and look at the tree through your legs. Can you see the top of the tree?

If you can't, move backward or forward until the top of the tree is visible. Mark the spot by scuffing the ground with your shoe, then use the tape measure to record the distance between the mark and the base of the tree.

The tree is the same height as this measurement. So, for example, if the tape measure records a distance of 20 ft (6 m), the tree is also 20 ft (6 m) tall.

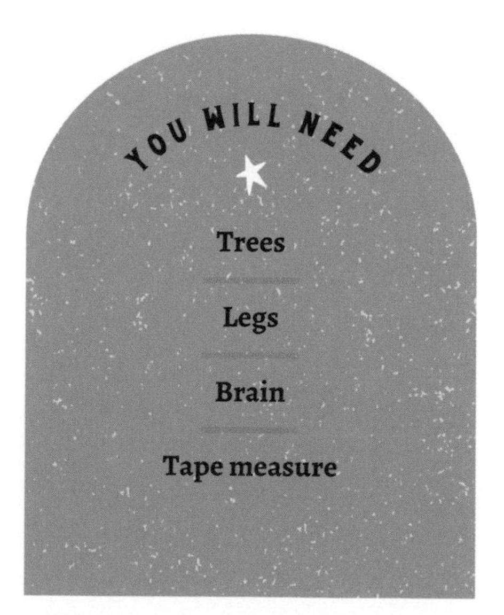

DID YOU KNOW?
The exact location of the world's tallest tree, which is a 380 ft (116 m) tall redwood called Hyperion, is kept secret to help limit visitors and keep it safe.

The method works because of a clever bit of math. When you can see the top of the tree through your legs, the angle between the ground and the top of the tree is approximately 45 degrees. Trees grow up in a straight line, so the angle between the ground and the tree trunk is 90 degrees.

Now think triangles. Imagine a triangle that goes from the base of the tree to its top, to your mark on the ground, and back to the base.

The angles in a triangle always add up to 180 degrees, so the third angle in the triangle must also be 45 degrees. This makes the triangle isosceles. Isosceles triangles have two equal sides, so this means that the distance between you and the tree is about the same as the height of the tree itself.

What is the height of the largest tree you can find? This method also works with other tall things such as houses, towers, and giraffes, so why not try measuring other tall objects? Just don't get too close to a giraffe!

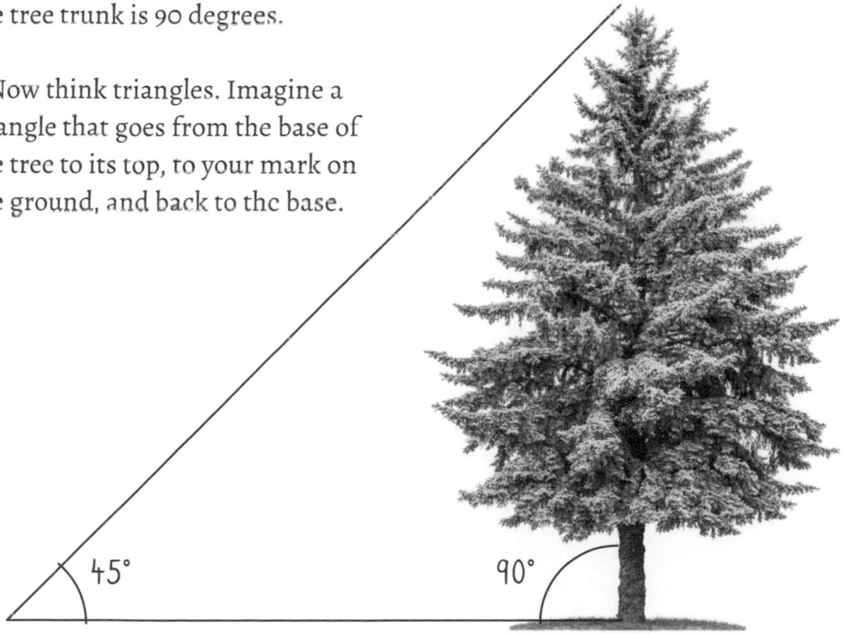

GROW A GRASSY MONSTER

Have you ever wondered what it would be like to have your very own pet monster? Here's a chance to find out. Create a monster with real, growing hair. Make it as ugly as you can.

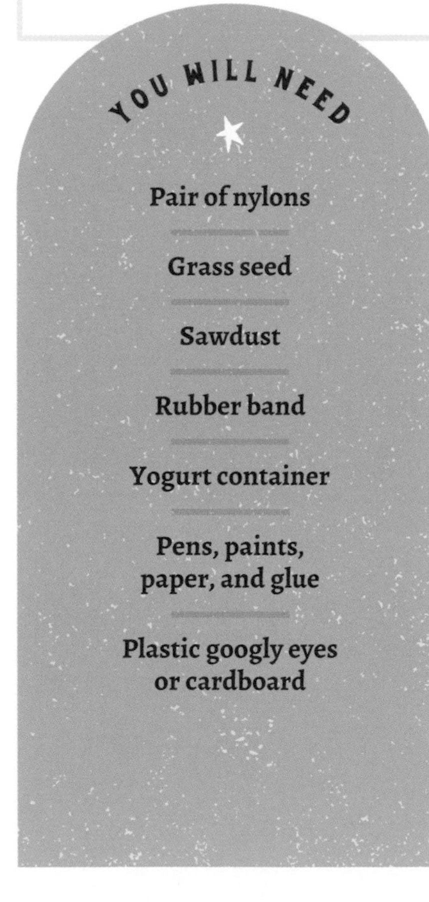

YOU WILL NEED

- Pair of nylons
- Grass seed
- Sawdust
- Rubber band
- Yogurt container
- Pens, paints, paper, and glue
- Plastic googly eyes or cardboard

Cut the foot off an old pair of nylons to make a short sock. Fill the toe of the sock with grass seed. Fill up the sock with sawdust so it forms a ball shape. This will be the monster's head. Monsters aren't pretty, so the head doesn't need to be perfect. If it's lumpy and bumpy, it will make your monster even more monstrous. Use the rubber band to close the sock shut so the monster's "brains" can't fall out.

Decorate the yogurt container. This will be the monster's body. Give the monster arms and legs, and maybe even a tail. What does the body look like? Is your monster covered in scales or spots? Is it wearing clothes or is it naked?

DID YOU KNOW?
The world's largest captive crocodile was more than 20 ft (6 m) long. What a monster!

Fill up the yogurt container with water and balance the monster's head on the body. Carefully glue some googly eyes and teeth onto the monster. If you don't have plastic googly eyes, you can make some eyes out of cardboard. How many eyes does your monster have? Just two? Some spiders have eight eyes, so you could add more.

Place your monster somewhere light and sunny, such as on a windowsill. Top off the water from time to time. The bottom of the monster's head should always be touching the water. The water will soak up through the sawdust and help the grass seeds to germinate.

A few weeks later, your monster should be sporting a fine head of spiky grass hair. Try styling the hair. Would your monster look good with pigtails, or should you trim the grass around the sides of the head and give it a Mohawk? Take a photo of your creation and put it in your journal. See how long the grass can grow, then when you are ready, cut it all off and start again.

LEAF SKELETONS

Humans have skeletons made of bone. Leaves have skeletons made of hollow vascular tubes. Boil away the fleshy parts of a leaf to reveal its intricate inner skeleton.

Leaves

Five cups of water

One cup of sodium carbonate

Old saucepan

Water

Bowl

Rubber gloves

Tweezers

Paper towels

Small, stiff brush

Choose some leaves. Thick, waxy, glossy leaves such as magnolia and holly work well, but take longer to prepare. Thinner, duller deciduous leaves such as beech and oak also work well, and are quicker to prepare.

Ask an adult to do the first part of this experiment for you. Add five cups of water and one cup of sodium carbonate to the saucepan. Sodium carbonate is also known as washing soda or soda crystals. It can irritate the skin, so be careful. Heat the mixture gently until the sodium carbonate has dissolved.

Using the tweezers, add the leaves to the mixture. Make sure they are fully immersed in the liquid. Allow the leaves to simmer for half an hour, until they are soft and bendy. If you are preparing waxy leaves, they will need to simmer for an extra thirty minutes.

- Prepare a bowl containing only water. Using the tweezers, transfer the leaves from the saucepan to the bowl. Let the leaves sit in the water for ten minutes. This softens the leaves further and helps to wash away the sodium carbonate.

- Remove the leaves from the water and place them on a piece of paper towel.

- Politely ask the adult to stop helping. Now it's your turn. Put on some rubber gloves to protect your skin. The leaves should be soft and starting to deteriorate. Use a small, stiff brush to tease away the fleshy parts of the leaf that surround the skeleton. This is delicate work.

Sometimes leaf skeletons occur naturally when microbes in the soil eat away at the fleshy parts of the leaf.

Be careful not to break the skeleton or the leaf. If the flesh doesn't come away easily, soak the leaf in water for another thirty minutes, then try again.

- Transfer the leaf to a new piece of paper towel and allow it to dry. You now have your leaf skeleton. Look at the intricate network of vessels that transport liquid around the leaf. Take a photo and put it in your journal.

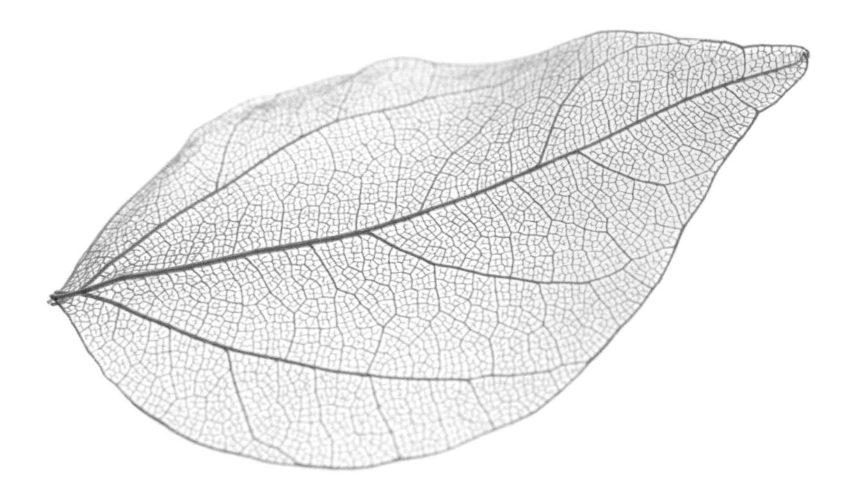

SEED DISPERSAL SPINNER

In fall some trees release "helicopter" seeds that spin through the air as they tumble to the ground. Make a paper seed spinner to show how these seeds are dispersed in the wild.

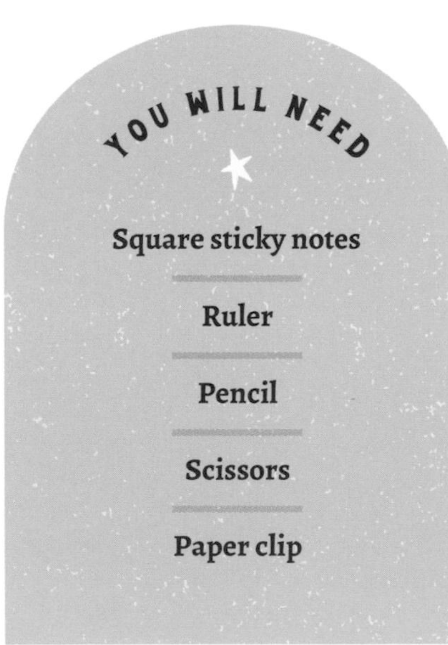

Place the sticky note on the table, sticky side up, and spin it around so the sticky strip is at the bottom. Draw a horizontal line three-quarters of the way across the sticky note. This is the line you will use for folding.

Make two equally spaced cuts that go from the top of the sticky note down to the folding line. This divides the top of the sticky note into three equal strips.

Fold the middle strip downward to make a "Y" shape. It should stick to the sticky part of the sticky note.

The outer strips can now be folded. Fold one strip toward you and the other strip away from you. This time the strips should not stick down. They should flap around freely. These are the "wings" of the helicopter.

The helicopter needs a little bit of weight, so slide a paper clip onto the middle, downward section of the "Y" shape. The seed disperser is now good to go.

Stand on a chair. Hold your arm up high, and let the helicopter go. How did it fly? What happens if you add more paper clips or change the shape of the wings? Make a note of your findings.

DID YOU KNOW?
The air pattern generated by a swirling helicopter seed is similar to the one made by hummingbirds when they swing their wings back and forth to hover.

Plants have evolved many ways to ensure that their seeds are dispersed as widely as possible. This is because if a seed germinates too close to the parent plant, the plants will have to compete with each other for resources such as light and nutrients.

Some seeds are dispersed by animals. Others are dispersed by the wind. Seeds that are transported by wind have some clever adaptations that help to keep them airborne. The wings on your seed disperser help to do this. As the seed disperser falls, the air pushing upward gives the seed lift. The two arms get a push in opposite directions. This makes it spin.

HOW OLD IS YOUR TREE?

Have you ever wondered how old the trees in your yard are? Most people know that you can estimate a tree's age by counting the number of tree rings it has, but to do this accurately, the tree has to be cut down. Try these alternate methods to estimate the age of a tree without damaging it.

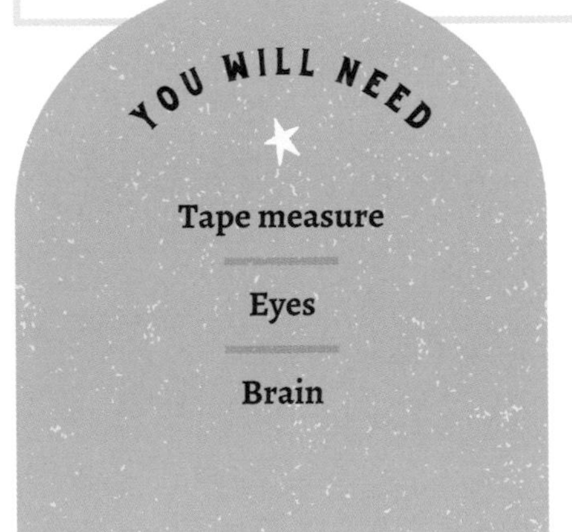

If the tree is in your yard or local park, do you or someone else know the year that it was planted? If you do, the task is simple. Subtract the year the tree was planted from the year it is now. This is the tree's age.

As trees grow, they lay down new layers of tissue around the trunk. This tissue forms the rings that some people use to age trees. Every year, as new rings are added, the circumference of the tree increases, so a tree's circumference can be used to help estimate its age.

Stand next to your tree. Measure 5 ft (1.5 m) up from the base of the tree trunk, then wrap the tape measure around the trunk and record the tree's circumference.

Trees grow at different rates. This depends on many factors, including the amount of rainfall and the availability of nutrients. In gardens and parks, the circumference of broad-leaved trees like oak and beech increases by around 3/4 in (2 cm) every year. To find a rough estimate of the age of your tree, simply divide its circumference by .75 (or 2 if measuring in centimeters).

If the tree is a conifer, there is another method you can use. Conifers are trees such as pines, cedars, and firs that produce cones. Unlike broad-leaved trees, many conifers grow in a regular pattern. Every year, a new set of branches grows out from the trunk. They're a bit like the spokes on a bicycle wheel. Whorls are the areas where the branches radiate out from the trunk, so if you count the number of whorls, you will have an idea of the tree's age. Take care not to count any small branches that are growing between the whorls. These are not genuine whorls.

What is the oldest tree you can find? The oldest known living tree is more than 4,850 years old. It's a Great Basin bristlecone pine that lives in the White Mountains of California. It's called Methuselah, after the long-lived character from the Bible.

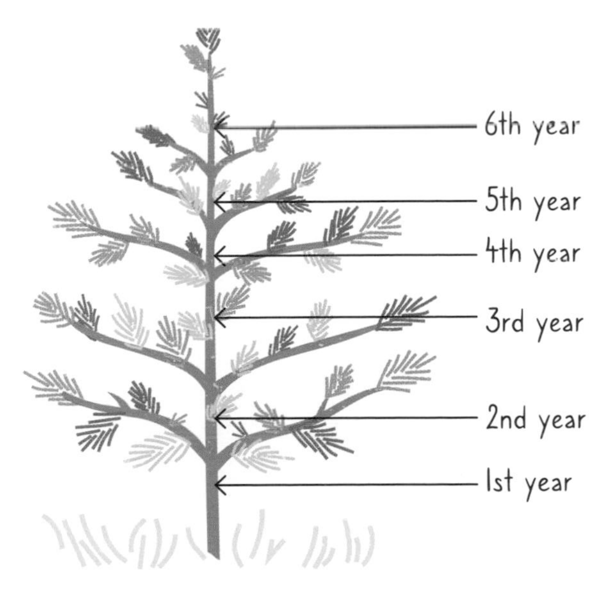

WHY DO LEAVES CHANGE COLOR?

In fall the leaves of many trees turn from green to gold, orange, and red. We're all familiar with this spectacular transformation, but have you ever wondered why it happens? Find out why leaves change color using this simple experiment.

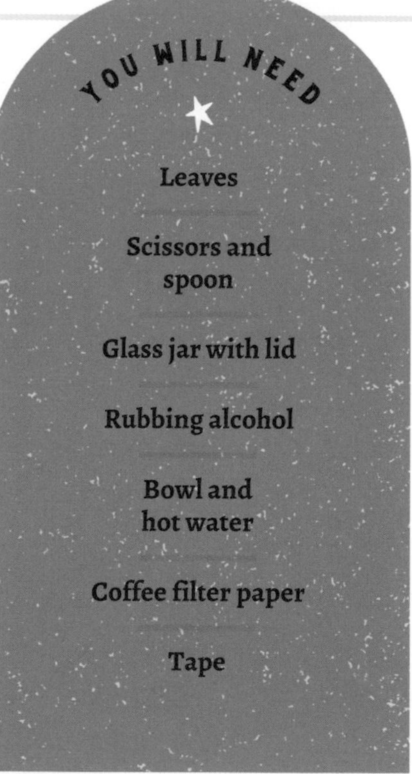

Collect a handful of green leaves. The leaves should come from a deciduous tree, such as an oak or a maple, which will shed its leaves when autumn comes. Using the scissors, cut the leaves into tiny pieces, then put them in the glass jar. Working outside or in a well-ventilated room, pour in just enough rubbing alcohol to cover the leaves, then mush the leaves up by pressing them against the sides of the jar with the spoon.

Put the lid on the jar. If you don't have a lid, cover it with tinfoil. Place the jar into a bowl full of hot tap water and leave it for an hour. Every ten minutes or so, pick up the jar and swirl it around. As the water cools down, replace it with more hot water. Do you notice anything happening to the liquid? The color should begin to change.

Cut some long thin strips from the coffee filter paper. The strips should be 1/2–1 in (1–2 cm) across and at least as long as the jar is tall. Take the jar out of the bowl and remove the lid or tinfoil. Dip one end of the strip into the leafy liquid and tape the other end to the outside of the jar so the paper can't fall in. Wait for a couple of hours. What do you see?

There should be at least two different stripes on your filter paper: one green and one yellow or orange. This shows you the different colors that are hiding inside the green leaves, just waiting to burst out in fall.

Chlorophyll is the green substance that makes plants green, but plants also contain yellow and orange substances that we can't see because they are masked by the green of the chlorophyll.

In fall, as the days become shorter, trees stop producing chlorophyll, and any chlorophyll that is left in the leaves is broken down. The green color fades, so the yellow and orange colors can finally be seen.

WATCH PLANTS BREATHE

Animals aren't the only things that breathe. Plants are living things, and they breathe too. Just like us, they "breathe in" one sort of gas and then "exhale" another. Watch this process happen with this simple experiment.

YOU WILL NEED

- Large glass bowl
- Water
- Large leaf
- Scissors
- Stone
- Magnifying glass (if you have one)

Fill the large glass bowl with warm water from a tap. Glass bowls are great for this experiment because they let you watch what's happening from all angles and slightly magnify the contents of the bowl. If you don't have a glass bowl, any bowl will do.

Head outside and find a leaf. It needs to be alive, so it should be fresh from a tree or other plant. The bigger the better. Choose the largest leaf you can find, then snip it at the stem and bring it inside.

Place the leaf inside the bowl of water and weigh it down with the stone. The leaf should be completely underwater. Place

DID YOU KNOW?
Plants are very important because they produce the oxygen that we breathe, and help to absorb the carbon dioxide that is produced when people burn fossil fuels such as oil and coal.

the bowl somewhere sunny. This could be outside in the sunshine, or inside on a windowsill. Now for the hardest bit. You need to wait. Let the experiment run for a couple of hours.

When you return to the experiment, take a look at the leaf. If you have a magnifying glass, use it to look at the leaf closely. What do you see?

The leaf should be covered in lots of tiny bubbles. Poke the leaf gently with your finger, and the bubbles should rise to the surface.

When we breathe, we breathe in a gas called oxygen, and breathe out a gas called carbon dioxide. Plants are different. They absorb carbon dioxide and water and convert it into sugars and oxygen. This is called photosynthesis, and it's powered by sunshine.

When we breathe, we use our lungs. When plants breathe, they use tiny holes in their leaves called stomata. The little bubbles that you can see are bubbles of oxygen that are escaping from the stomata in the leaves.

HOW TO GROW PLANTS IN SPACE

In the future, when humans travel to Mars or other faraway planets, they'll need to be able to grow their own food—but soil is heavy and floats around in space! So, scientists are planning ahead and growing plants without soil. This is called hydroponics. Grow your own space plants by sowing seeds soil-free.

YOU WILL NEED

- **Large plastic bottle**
- **Scissors and string**
- **Cotton balls**
- **Water**
- **Bean seeds**
- **Modeling clay**
- **Wooden skewers**

Remove the lid from the large plastic bottle. Using scissors, carefully cut the bottle in half. Take the top half and turn it upside down, then place it into the bottom half of the bottle. This is where the seeds will go.

Using the top part as a funnel, fill the bottom part of the bottle with water. The water level should be just below the upside-down neck of the bottle.

Cut your string into five pieces. They should be at least three times as long as the height of your hydroponic container. Feed the pieces of string down through the neck of the bottle so they dangle in the water. Don't push too much string through the neck of the bottle or it will fall through. Most of the string will be all tangled up inside the upside-down top half of the bottle. This is just fine.

- Put five or six cotton balls on top of the tangled string and sprinkle a handful of bean seeds on top of the cotton balls.

- Make a tripod to support the growing seeds. This is a bit tricky. Make three little blobs of modeling clay and arrange them into a small triangle shape on a table. Press one skewer into each blob, then use another small piece of string to tie the other ends of the skewers together. The structure should look like a little tepee. Now balance the tripod on top of the cotton balls and move the container into a bright sunny place. Watch to see what happens.

- The water will soak up through the strings and into the cotton wool. After a few days, the seeds will germinate. The roots will grow down through the neck of the bottle into the water, and the shoots will grow up and around the tripod.

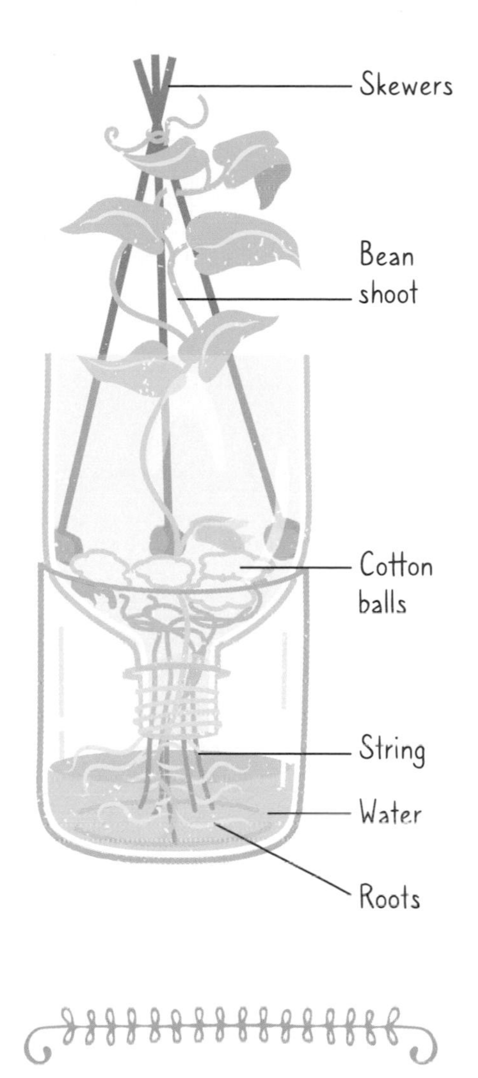

After a while the seedlings will need extra nutrients. You can buy hydroponic food online, but if you have a fish tank, pond, or nearby lake, you could try using some of the water from that.

MAKE A MINI GREENHOUSE

Greenhouses help plants to thrive, but just how much of a difference do they really make? Make a mini greenhouse from old CDs and see how temperature affects plant growth.

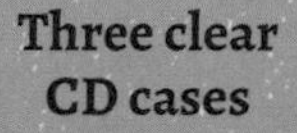

YOU WILL NEED

- **Three clear CD cases**
- **Clear tape**
- **Sweet pepper seeds**
- **Two small plant pots**
- **Soil**
- **Thermometer and ruler**

Open up the CD cases and remove the plastic tray that usually holds the CD. This can be recycled with the plastic waste. The first two CD cases will form the four walls of the greenhouse. Open up the cases so the hinges are opened to ninety degrees. Stand the two cases up and arrange them so they form a box. Stick the cases together using tape. At this stage, the box has no bottom and no top.

The third CD case is going to form the roof of the mini greenhouse and reinforce one of the walls. Position it so that one side of the case rests on one of the walls to form an additional wall panel (it's a bit like having double panes in a window), and the other side of the case forms a horizontal roof. Stick the new wall panel to the structure using tape, but take care not to stick the roof down. Gardeners ventilate their greenhouses by letting a little air in. This will enable you to do the same.

★ The greenhouse is now ready to go. Fill the two pots with soil. Add a couple of sweet pepper seeds to each of the pots and sprinkle a little extra soil on top. Water the seeds so the soil becomes moist. Put the two pots on a windowsill but cover one with the mini greenhouse. Place the thermometer inside the greenhouse.

★ Now watch and wait for the peppers to grow. Give the seedlings a little water every other day, or whenever the soil is close to drying out. Which seedling germinates first? When the seeds have germinated, measure how quickly they grow. Use a ruler to measure their height twice a week. Measure the temperature inside the greenhouse, then remove the thermometer and measure the temperature outside the greenhouse. Record your observations in your science journal.

Greenhouses keep plants warm, even in the depths of winter. The glass (or plastic) traps the sun's heat so sometimes the temperature inside the greenhouse can be more than 18°F (10°C) warmer than the outside temperature.

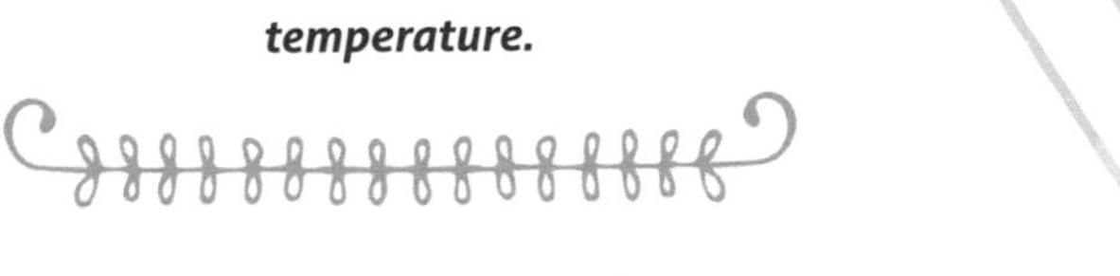

MAKE AN ECOSYSTEM IN A JAR

Plants use the carbon dioxide we exhale to help make the oxygen that we breathe. Demonstrate some of the ecological cycles that keep us alive by creating an ecosystem in a jar. See how long you can make the ecosystem last.

YOU WILL NEED

- Large glass jar
- Pebbles and charcoal
- Soil and potting compost
- Plants
- Spoon
- Water

Ecosystems are communities of living things and a place where they live. For example, your garden is an ecosystem because it contains living things such as birds, insects, bacteria, and plants, and physical structures such as fences, stones, and water.

To build an ecosystem in a jar, first put a layer of pebbles in the base of the jar. This will give any excess water somewhere to collect, so the plants don't drown!

Sprinkle a thin layer of charcoal over the pebbles. You can buy charcoal from a store or collect it from the ashes of a disused barbecue or bonfire. The charcoal is important because it acts as a filter, helping to collect impurities and keep the ecosystem healthy.

* Prepare a mixture that is half soil and half potting compost, and add a thick layer to the jar. Potting compost is good because it contains plenty of nutrients to help the plants grow, and the soil is important because it's packed with bacteria. Together, the three layers should fill about a third of the jar.

* Add some small plants. Anything will do, but choose plants that require a similar amount of water. So, for example, a cactus and a daisy will not go well together. The plants should easily fit in the jar with room for growth. Dig a small hole in the soil with a spoon and add the plants. Cover the roots with soil.

* As a finishing touch, you could add a bigger pebble in among the plants. There's no real reason for this. It just looks professional! Now put the lid on the jar and place it somewhere that is well lit.

* The idea is that the ecosystem will look after itself, so there's no need to water it or add in air, but for the first couple of days it may need some assistance. If the soil looks dry, add a little water. If the inside of the jar is always covered in condensation, then open the lid to let it breathe for a while.

Take a photo of your ecosystem and stick it in your journal. Record how long the ecosystem lasts before the plants start to die.

SMART PLANTS

Some people talk to their plants because they think it helps the plants to grow, but did you know that plants are very good at communicating? They may not have mouths to talk with, or ears to hear with, but plants are constantly sending out and receiving information.

If a flowering plant wants to attract the attention of a pollinating insect, for example, it may release chemicals into the air. There is a plant that grows in Sumatra called the titan arum, or "corpse flower," (right). It releases a chemical called dimethyl trisulfide, which smells like rotting flesh. Insects that are flesh-eating, such as some beetles and flies, find the smell irresistible, so they visit the flower, pick up its pollen, and then spread it around the neighborhood when they finally buzz off. The corpse flower might not talk out loud, but it is still communicating with a different species.

★ Plants can also communicate with one another. When bean plants are attacked by leaf-eating flies, they release chemicals into the air that warn the plants' neighbors of an impending attack. In return, the neighbors release different chemicals that repel plant-eating flies and attract fly-eating wasps.

★ Plants also communicate with one another underground. If one plant is struggling because it has no water, it sends a message via its roots, to plants that are nearby. The neighbors respond by closing tiny pores, called stomata, that normally allow water to evaporate from their leaves. This helps the plant to retain the water that it already has and prepares it for future drought.

★ The more researchers study plants, the more they realize that plants have complex social lives. Trees, for example, can recognize plants that are closely related to them. Sometimes they grow their roots toward their young so they can transfer sugar to them and help them grow. Other times they stop growing their roots toward their young to prevent competition between "parent" and "child."

★ Plants are social and plants are smart. Did you know that some plants can even count? The Venus flytrap (above) catches insects inside modified leaves that can quickly snap shut. The leaves have specialized trigger hairs that sense the insects, but the trap only shuts if the hairs are touched twice within a period of ten seconds or so. It then takes a further three touches, when the insect is struggling, for the plant to start producing enzymes that digest the insect. So, the Venus flytrap is doubly smart because it can count at least two things: time and the number of touches.

0.

KITCHEN SINK SCIENCE

It's great being in the backyard or garden, but sometimes it's hard to get outside. When it's rainy or cold, you can bring the joys of the outside world into the home by performing some experiments indoors. Kitchen sink science is all about creativity and innovation, and it is inspired by the natural world. Who says you need to leave the house to enjoy the great outdoors?

KITCHEN SINK SCIENCE

The experiments in this section draw their inspiration from the garden and the natural world. They all use natural items, such as potatoes, sticks, and lavender, which can either be grown or found in the garden. Some of the more exotic ingredients, such as avocados and lemons, can be bought from a supermarket or borrowed from a fruit bowl.

Most of the experiments aren't actually done in the kitchen sink (one of them is), but some of them do involve washing dishes. My children try to persuade me that doing the dishes is an adult's job, so you may want to see if you can pull off the same scam!

The experiments in this section are creative and diverse. In this part of the book, you'll be making lots of "ometers." There's a barometer to predict the weather, a thermometer to record the temperature, and an anemometer to measure wind speed. After that, the national weather forecasters had better watch out, as you could be after their jobs.

Go one step further and you could even make some weather. There's a really great experiment to make a cloud in a jar, and another to make a tornado in a bottle. Best of all, you won't get soaked, and along the way you'll learn a little about how the weather works.

There are lots of other fun things to try. Decorate your windows with cardboard silhouettes to stop birds from flying into them. Make a pond-dipping net so you can go outside when the rain stops and have an adventure.

Liven up your garden walls (literally) by making a living, breathing, moss-based paint that you can use outside. Eco graffiti, as it's called, is nontoxic and environmentally friendly, and if you're any good at art—which I'm sure you are—the results will be lovely to look at.

By the end of this section, you'll be washing with horse chestnut soap, bathing in homemade bath bombs, and relaxing with scented lavender bags. Experiment with the natural ingredients that surround you and see if you can come up with your own kitchen sink science experiments.

TURN POTATOES INTO SLIME

Potatoes may seem ordinary, but they're anything but. Learn about weird materials by making a spud-based substance that is solid when you play with it and liquid when you leave it alone.

Grab a bag full of potatoes and give them a wash. Cut them up as small as possible and place the pieces in a large bowl. Cover the potato pieces in hot water and use the wooden spoon to stir the mixture for five minutes. Keep the mixture moving. Do not let the water stand still. What do you see? The water may start to change color. This is normal.

Acting quickly, strain the potato mixture through the sieve, pouring the liquid into a second large bowl. You don't need the potato chunks now, but it's a shame to throw them away, so ask an adult to turn them into something delicious.

Meanwhile, allow the freshly sieved potato water to settle in the new bowl. Do not stir it or disturb it in any way. Leave it for at least fifteen minutes. Now what do you see? The mixture should start to separate.

Pour away the upper layer, leaving the thick, white lower layer behind. This is your slime, but it's still a bit dirty so it needs to be washed.

The potato slime isn't a solid, but it's not a liquid either. It's something called a non-Newtonian fluid. Non-Newtonian fluids are weird. They go solid when they are under pressure, but they liquefy when the pressure is removed.

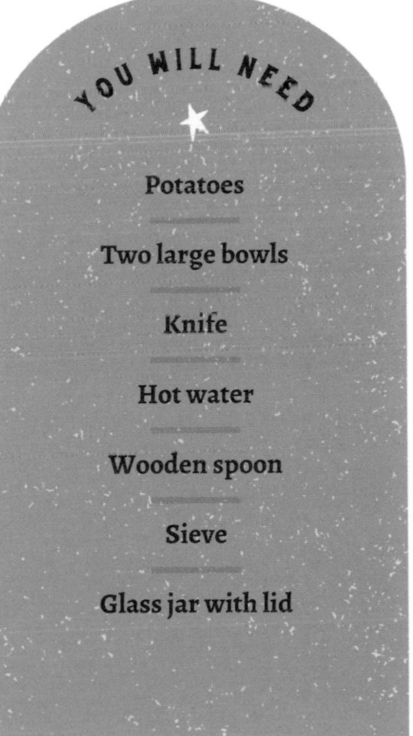

Mix the slime with cold water and transfer it to a glass jar. Put the lid on and shake the mixture vigorously. Leave it to settle for ten minutes. The white bottom layer will reappear. Quickly pour away the top layer of liquid, and there you have it: your very own potato slime.

Take the milky mixture out and knead it with your fingers. The more you manipulate it, the more doughlike it should become. Try punching it and the mixture should resist. Now what happens when you stop handling it? The semisolid mixture should "melt." It will turn into a liquid and slide through your fingers. Messy, isn't it?

MAKE RED CABBAGE pH PAPER

Red cabbage is not only good for you, but it also contains a natural pH indicator that changes color depending on the acidity of the surroundings. Make your own pH paper strips, then test the pH of some liquids from your home and backyard.

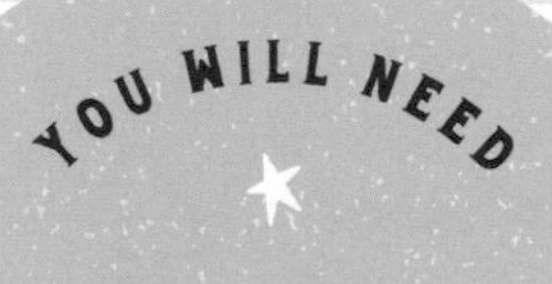

Half a red cabbage

Sharp knife and scissors

Boiling water

Heat-proof bowl

Jug or jar

Sieve

Coffee filter paper

Household liquids, e.g. vinegar or laundry detergent

Using the knife, chop up half a red cabbage into very small pieces. Place it into the heat-proof bowl and cover it completely with freshly boiled water. Leave the mixture for fifteen minutes. What do you see? The water should begin to change color as colorful molecules called pigments leak out of the cabbage.

Sieve the mixture to separate the cabbage from the water. It's a shame to waste good food, so ask an adult to cook the cabbage, or give it to a guinea pig! Collect the red-purple liquid in a jug or jar. This is your pH indicator.

Remember, pH is measured on a scale of 1 to 14. Liquids with a pH of less than 7 are acidic. Liquids with a pH of more than 7 are alkaline. Your pH indicator will have a neutral pH of about 7. Its exact color will depend on the pH of the water you used.

Prepare the pH paper. Cut the coffee filter paper into small strips that are about 2 in (5 cm) long and 1/2 in (1 cm) wide. Place the strips into the pH indicator and leave them to soak. The strips should be completely immersed in the liquid.

After a few hours, remove the strips and allow them to dry. You can hang them on a clothes line or put them on a heating vent or windowsill. Be careful—the color can run and cause staining.

Now it's time to get testing. Collect the samples that you want to analyze. Vinegar, lemon juice, and laundry detergent all give a good response. Place a drop of the sample onto the pH paper. If the paper becomes a blue, green, or yellow color, the sample is alkaline. If the paper becomes a dark red or purple, the sample is acidic. Record the results in your journal.

You can work out the pH of your sample by comparing the color of the pH paper against the color chart on this page.

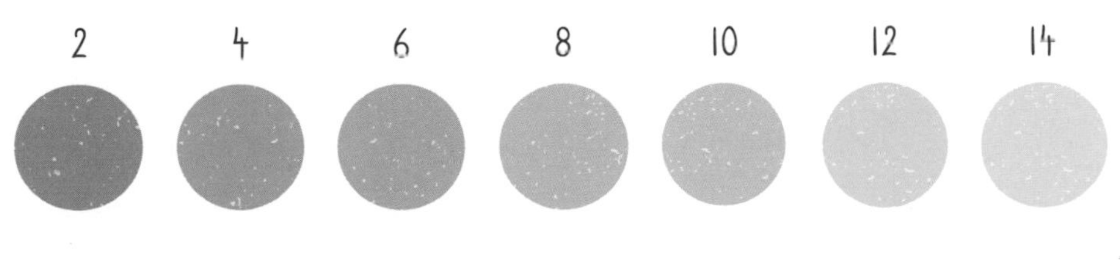

DYE IT WITH AVOCADO

Kitchens and gardens are full of plants containing natural chemicals that can be used to dye fabrics. Make a dye from avocado pits, then see if you can make different-colored dyes from different plants.

Four avocado pits

Plastic bag

Wooden cutting board

Hammer

Old saucepan and water

Sieve and wooden spoon

Old blouse or T-shirt, pillowcase or sheet

Put the avocado pits inside the plastic bag and place it on the wooden cutting board. Take the hammer and give the pits a gentle whack. The idea is to break the pits in half and then smash them into smaller pieces. The bag is there for protection as it stops bits of avocado pits from flying all over the place.

Put the small, freshly crushed avocado pieces into an old saucepan and add 1.8 pints (1 liter) of water. Heat the mixture until it begins to boil, then reduce the heat and let it simmer gently for an hour. Ask an adult for help. This will help to draw the dye out of the avocado pit and concentrate the dye. Make sure

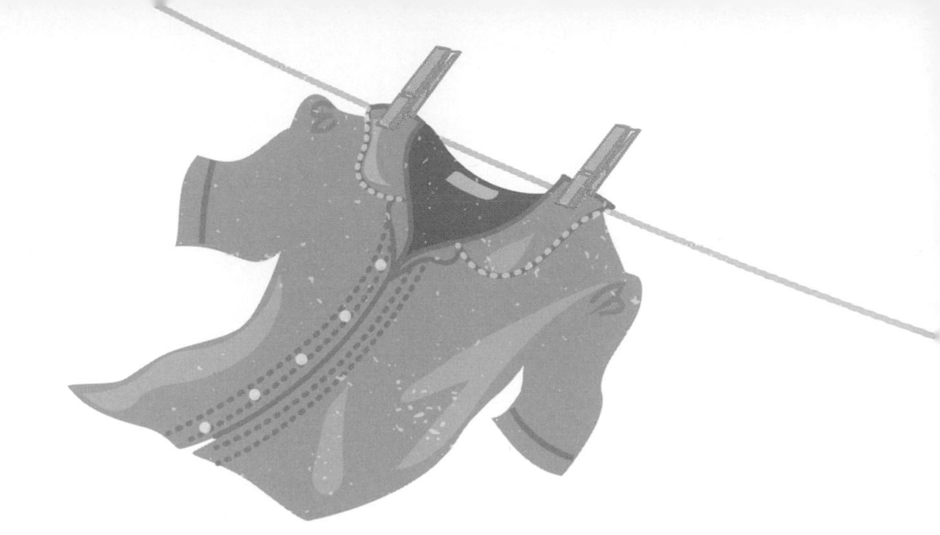

the pan doesn't boil dry. Top it off with a little water if needed. Strain the mixture through a sieve and return the liquid to the saucepan.

Prepare the fabric. White cotton works well, so you could use an old blouse or T-shirt, or cut up an old pillowcase or sheet (check with an adult first). Moisten the fabric by running it under a tap, then add it to the saucepan. Make sure the fabric is completely covered. Swirl it around with a spoon and leave it overnight.

The next day, remove the fabric and wring it out. Leave it to dry on a clothesline or heating vent. When it is dry, ask an adult to iron the fabric on a medium heat. This will help to fix the color so it won't run when the fabric gets wet.

DID YOU KNOW?
You can also make ink from avocado pits. Try using some of the leftover dye to paint a picture.

Avocado dye stains fabric a light, rosy pink color, but what other colors can you make using the natural ingredients that are around you? Experiment with different items from the garden and the kitchen. Onion skins, carrot tops, coffee grounds, blueberries, nettles, and pomegranate seeds also make excellent dyes. In your journal, record which items work best.

HELP BIRDS AVOID WINDOWS

Birds sometimes fly into windows and injure or kill themselves. Become a bird hero and help prevent this from happening by making your windows bird-proof. The birds will thank you for it!

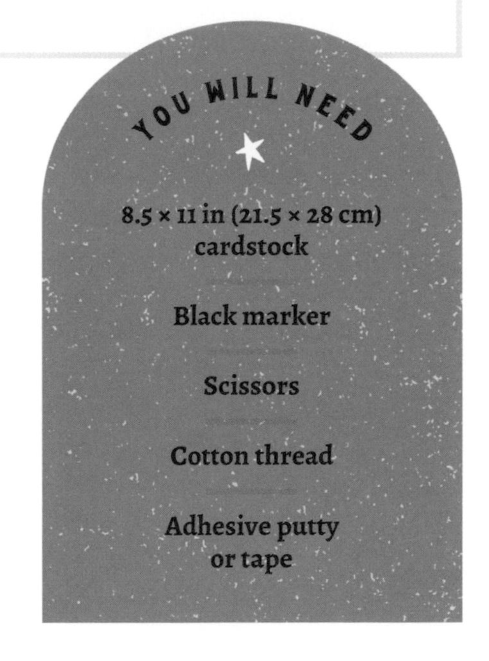

Birds sometimes fly into windows or glass doors because they simply don't see them. The glass acts like a mirror, reflecting images of the trees and the sky, so often the birds don't realize that the glass is actually there.

To prevent birds from flying into windows, the panes need to be made more obvious. This can be achieved simply by hanging objects in front of them.

Small birds are frightened of large predatory birds, such as hawks, and will go to great lengths to avoid them. Sketch a picture of a hawk on a piece of card. Its wings should be outstretched as if in flight. It doesn't need to be a masterpiece, but if you find this difficult, you can print out a picture from the Internet and use it as a template.

Color the hawk in using the black marker. The idea is to create a dark silhouette that will be easy for garden birds to spot. Using the scissors, cut the hawk out. Make a small hole in the bird's body and attach a long piece of cotton thread. Hang the bird in your window using adhesive putty or tape.

Repeat the process. If the window is particularly big, add more than one hawk. Try making a silhouette scene, including trees, flowers, clouds, and a sun. Birds are less likely to fly into a window filled with lots of obstacles. Make a note of what happens over time. Is your bird deterrent successful?

DID YOU KNOW?
In some countries, a group of hawks is called a "kettle" of hawks, but in reality, it's rare to see hawks together. Adult hawks tend to be solitary birds that only come together when they breed or migrate.

This simple measure should greatly reduce the number of collisions that occur, but if you do find a bird that has flown into a window, treat it with care. It may be suffering from concussion or have internal injuries. Place it in a quiet, dark place and leave it for a couple of hours. Hopefully, the bird will recover.

AMAZING MOLD EXPERIMENT

The world is full of living things so small they can't be seen with the naked eye. When they grow in vast numbers, they become visible. Watch microscopic mold grow into something big on a slice of bread and learn about the conditions needed to make it grow.

YOU WILL NEED

- Three slices of white bread
- Three pieces of paper towel
- Water
- Three clear plastic bags
- Masking tape

If you've ever opened the bread bin and found that the loaf is covered in colorful fuzz, that's mold. Mold is a type of fungus. It grows from spores that blow around in the air, and can be found just about everywhere, including on your skin, in the soil, and on work surfaces. Some molds can make people ill, but most are totally harmless.

Prepare your mold-growing bags. Fold the pieces of paper towel into quarters and sprinkle them with water so they are damp but not soaking. Now put one piece of paper towel into each of the plastic bags.

Press your hand firmly into each slice of bread so it leaves a handprint in the middle. Some of the spores on your skin will be transferred to the bread. Put one piece of bread into each of the bags so it sits on top of the paper towel. Now seal the bags tightly shut using the masking tape.

Label the bags 1, 2, 3. Put the first bag in the fridge. Put the second bag onto a warm, brightly lit windowsill. Put the third bag into a dark kitchen cupboard. Now leave your mold to grow.

Check on the experiment every day. You should see mold beginning to form after a couple of days. After seven to ten days the experiment will be done. Do not open the bag. You can study the mold from the outside.

DID YOU KNOW?
Some molds can kill bacteria. Penicillin is a common antibiotic that was originally discovered in mold.

Do all of the bags contain moldy bread? Which contains the most mold? Has it grown in the shape of a hand? What conditions does mold need to grow? Take photos and describe your results in your science journal. When you are finished, ask an adult to dispose of the experiment in an outside trash can.

MAKE A POND DIPPING NET

Pond dipping is a great way to learn about all the creatures that live in your neighborhood pool of water, but it's hard to do without a net. Make your own pond dipping net, then go out and survey some aquatic animals.

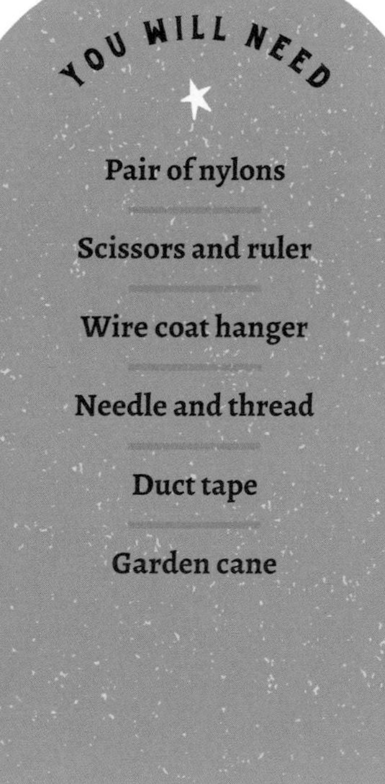

Nylons are thin and stretchy. They let water pass through but successfully retain small swimming creatures that would otherwise get away. Make sure your adult has finished with the pair of nylons you are going to use since they will never be wearable again.

Measure down about 6 in (15 cm) from the crotch of the nylons. The crotch is the bit between the two legs. Cut off the legs of the nylons at this point. Throw the sock bits away or use them to make a bug sucker (see pages 36–7). Gather what's left of the two legs together and tie them into a tight knot. This is going to form the main part of the net.

Open up the wire coat hanger so it makes a diamond shape. Place the net in the middle of the diamond and fold the waistband over the metal. Fasten the nylons to the coat hanger by sewing all the way around. A simple running stitch is good for this. If you need some help, ask an adult.

See how many different species you can catch with your pond net or swoop net. Record them in your journal.

Straighten the coat hanger hook and secure it to the end of the garden cane using duct tape. Your pond dipping net is ready to go. Check out pages 40–1 to learn how to go pond dipping. If you're lucky, you might catch a newt like the one pictured above.

If you use an old pillowcase instead of a pair of nylons, you'll be able to make a swoop net, which can be used to catch butterflies. Swoop nets tend to be deeper than pond dipping nets because the butterflies will easily fly out of nets that are too shallow. Swoop nets are so called because you need to swoop them through the air in order to catch the butterfly.

MAKE A LAVENDER BAG

Lavender contains aromatic oils that have a relaxing effect. Make a mini lavender bag to help you sleep, then experiment with other herbs to see if you can improve on the scent.

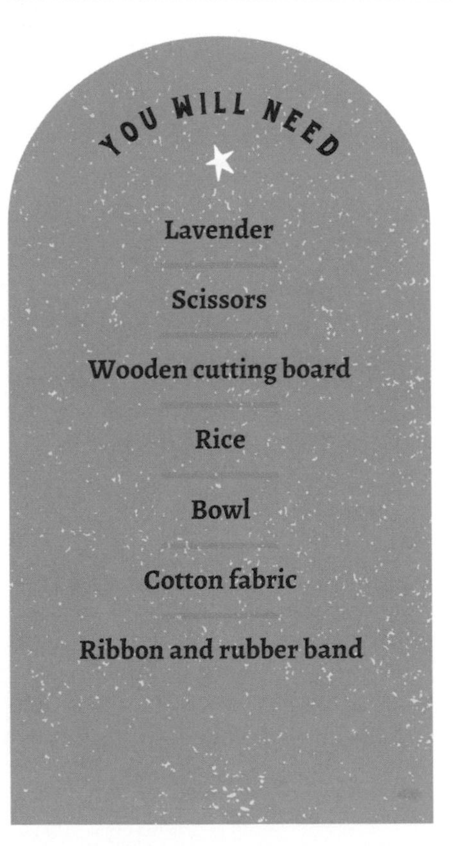

Prepare your lavender. Lavender plants flower in the summer. Harvest the flowers as they come into bloom. This is when their scent is strongest. Cut the flowers at the base of the stems then lay them out on a wooden cutting board. Put the cut lavender in a sunny spot, such as in a conservatory or on a windowsill. Leave it for at least a week until the lavender has completely dried out.

You may notice that the flowers begin to fade and lose their brilliant purple color. This doesn't matter. It's the scent that is important. Check every few days to see if the lavender is dry. The flowers will start to crumble from the stem when they are ready.

Prepare the stuffing for the bag. In a bowl, mix two parts of lavender to one part of rice. The rice helps bulk out the bag and make it less flimsy.

Prepare the fabric for the sachet. Cut a circle out of the material. The circle should be about the same size as a CD. Place the fabric circle—design-side down—on a table, then add a handful of the lavender mixture to the middle. Carefully gather up the sides of the circle to make a pouch or bag, then use the rubber band to fasten the bag shut. Tie a colorful ribbon around the rubber band to make the lavender bag look more attractive and professional. Put the bag next to your pillow at night, or use it to freshen up a closet or drawer. It's so much better than the smell of socks!

Lavender smells sweet but so do many other plants. Some geranium leaves smell of lemon, and herbs such as rosemary and thyme also have pleasant scents. Try making some alternate scent bags using other dried garden plants. Make a note of which smell you like the best.

Lavender plants are covered in oil glands. These are tiny star-shaped hairs that can be found on the leaves, flowers, and stems. When you rub the plant between your fingers, the glands release oil, which gives the plant its smell.

POTATO POWER

Make a potato battery and use it to power a calculator. See how long the battery lasts, then try making more powerful batteries from other vegetables and fruits.

YOU WILL NEED

- Potato
- Small piece of copper wire
- Galvanized nail (one that is covered in zinc)
- Old, inexpensive calculator
- Pliers and scissors
- Two electrical wires with clips at the end

It's easy to make a potato battery. Bend the copper wire in half and poke the sharp ends into the potato. Touch your tongue on the copper wire. Do you notice anything?

Poke the galvanized nail into the potato. The nail and the wire should be close to each other but not touching. Now touch the copper wire and the nail with your tongue. What do you notice?

When your tongue touched just the copper wire, you probably didn't notice much, but this time, you might feel a tingle or notice a metallic taste. This is a sign that your battery is working. Potato batteries generate a very small electric current that is harmless, but don't go licking store-bought batteries as these are more powerful and you may receive a small electric shock.

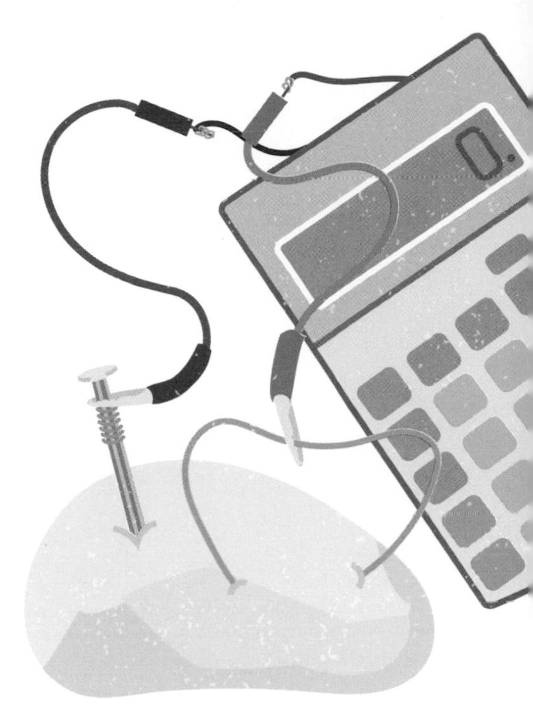

- Remove the battery casing from the back of your calculator. Warning! Your calculator may never be the same again so make sure it's an old, inexpensive one that you don't mind destroying!

- Pop out the battery and, using pliers, carefully detach the red and black wires that are attached to it. If your calculator has a solar panel, this will also need to be disconnected. Cut the wires that connect the solar panel to the calculator.

Now make some batteries using other kitchen items. Citrus fruits, such as oranges and lemons, work well, as do dill pickles and pickled onions. Which battery powers your calculator for the longest?

- Connect the battery to the calculator. Using one of the leads, clip the black, negative calculator wire to the nail. Using the other lead, clip the red, positive calculator wire to the copper wire. Turn the calculator over and look at the display screen. Your calculator should now be working.

- The battery works because electrons, which are tiny, negatively charged particles, are moving around the circuit you have created. The potato is an "electrolyte." Electrolytes are substances that help keep electric currents flowing.

ECO-GRAFFITI

Many people don't like normal graffiti because it's done with paint and is hard to remove. Eco-graffiti is made with living, growing plants. It's environmentally friendly and easy to remove. Make some with moss and literally bring a wall to life.

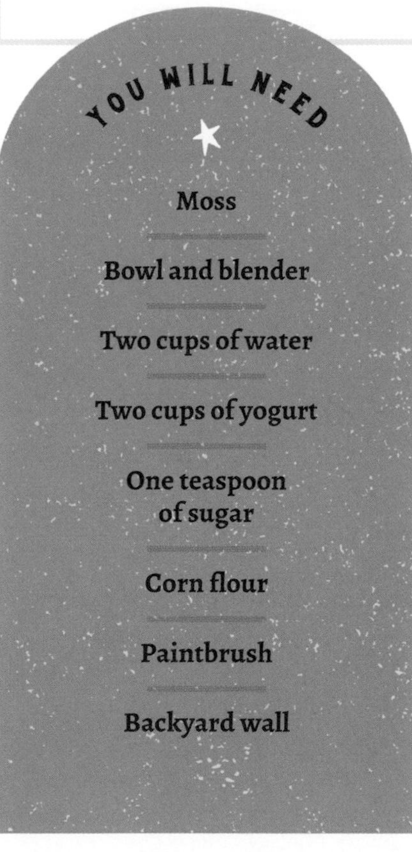

This is an activity to prepare in the kitchen and then do outside. First, collect a big handful of moss. Mosses are small flowerless plants that grow in thick green clumps in shady places such as forest floors or tree trunks. There are around 12,000 different species. They have thin simple leaves that are packed full of a green substance called chlorophyll.

Put the moss in a bowl and fill it with water. Give the moss a cleaning by swirling it around with your fingers. This will get rid of any soil that is still attached to it. Remove the moss from the water and break it into small chunks.

DID YOU KNOW?
Mosses don't have roots, stems, or flowers. They attach themselves to surfaces using tiny filaments called rhizoids.

Place the moss in the blender. Add the water, yogurt, and sugar. Blitz the mixture until it is completely smooth. You might want to ask an adult to help you with this. Remarkably, this doesn't kill the moss. It just breaks the plant up into lots of tiny pieces.

The mixture should have a paint-like texture. If it's too thick, add a little water. Add a little corn flour if it's too runny. The moss paint is now ready to use.

Ask an adult if you can try your eco-graffiti on an outside surface, such as a wall or the side of a shed. Remember, moss grows best in shady places, so avoid painting your design onto a surface that is in full sun. Use a brush to paint your design onto the wall. Why not paint something environmental, such as a tree or an animal?

Check the eco graffiti every week. If it takes a while to get going, you may need to add some more paint. If it looks like it's drying out, spray it with water. Take a photo and stick it in your journal. How long can you keep the eco-graffiti alive for? Try painting on different surfaces such as brick and wood. Which surface is best for growing eco-graffiti?

MEASURE WIND SPEED WITH AN ANEMOMETER

Some days there is hardly any wind at all, while other days the wind is so strong it feels like it could blow you off your feet. Build a device to measure wind speed, called an anemometer, then see how it changes from day to day.

Using the hole punch, make two holes opposite each other just below the rim of one of the paper cups. Push one of the straws through these holes so it sticks out on either side.

Now turn the cup through 90 degrees and repeat the first step. The cup should now have two straws poking through it, and they should cross in the middle. This is the base.

Calculate the wind speed. Hold the anemometer out of a car window while being driven along at 10 mph (16 km/h). Count the number of rotations in thirty seconds. If you count five rotations, you know that five rotations equals 10 mph (16 km/h). This is your benchmark. When you use the anemometer later, you'll know, for example, that ten rotations in thirty seconds means a wind speed of 20 mph (32 km/h).

Take another paper cup. Using the hole punch, make two adjacent holes halfway down the cup. The holes should be roughly 3/4 in (2 cm) apart. Do the same for the remaining three cups. These are the sails.

Now take these sails, one at a time, and push them onto the straws sticking out from the base. Guide the straws through the holes in the sails so they just poke out the other side. The sails should all be facing in the same direction.

Use the tip of a sharp pencil to poke a hole in the bottom of the base. The hole should be in the center of the base, and it should be big enough to poke a pencil through. Now push the blunt end of the pencil through the hole so the eraser rests underneath the place where the two straws cross.

Holding the pencil for support, lightly push the thumbtack through the two straws and into the eraser. Be gentle! If you press it in too hard, the anemometer won't work. Try giving the anemometer a spin. If it rotates freely, it's good to go.

Stand outside on a windy day and count how many times the anemometer spins in thirty seconds.

YOU WILL NEED

Five small paper cups

Two straws

Ruler and pencil with an eraser on the end

Thumbtack and a hole punch

An adult with a car

MAKE A ROSE PETAL BATH BOMB

Lie back and relax with some homemade bath bombs scented with ingredients from the garden. Which flowers give the most satisfying scent?

YOU WILL NEED

- 10½ oz (300 g) baking soda
- 5 oz (150 g) cream of tartar
- Two teaspoons olive oil
- Large mixing bowl
- Essential oils (any you like)
- Wooden spoon
- Food coloring
- Handful of rose petals
- Water in a spray bottle
- Silicon ice-cube tray

★ Add the baking soda, cream of tartar, and olive oil to a large mixing bowl. The olive oil helps to bind the ingredients together and will moisturize your skin in the bath. Add a few drops of essential oil. Any will do, but lavender oil works really well.

★ Use a wooden spoon to mix all of the ingredients together. Add a few drops of food coloring. Do not add the whole bottle. You probably don't want to end up with blue or pink or green skin.

★ At this stage the mixture will still be powdery. This is to be expected. Tear up the rose petals into small pieces and add them to the mix. Give it another stir, then add a few sprays of water. Listen carefully.

What do you hear? The mixture should begin to fizz as the ingredients start to react.

Keep stirring the mixture until it becomes the texture of wet sand. Don't add too much water or all the fizz will disappear and there will be none left for the bath.

Take small handful-sized blobs of the mixture and press them into the silicon ice-cube mold. The mold will help to stop the bombs from crumbling. Take care not to touch your eyes at this point, because the mixture could irritate them.

Rose petals have a beautiful scent, but what other plants can you find that have soothing aromas? Try making bath bombs with crushed lavender seeds or grated orange peel.

Now for the difficult part. Leave the bombs to set. This will take one to two days. Make sure they are totally dry and hard before removing them from the mold. Add them to a bath full of warm water, step in, and relax.

TORNADO IN A BOTTLE

When they occur naturally, tornadoes can destroy trees, houses, and cars. Make a mini tornado in a bottle. It won't destroy anything (I promise), but it will help to explain how real tornadoes form.

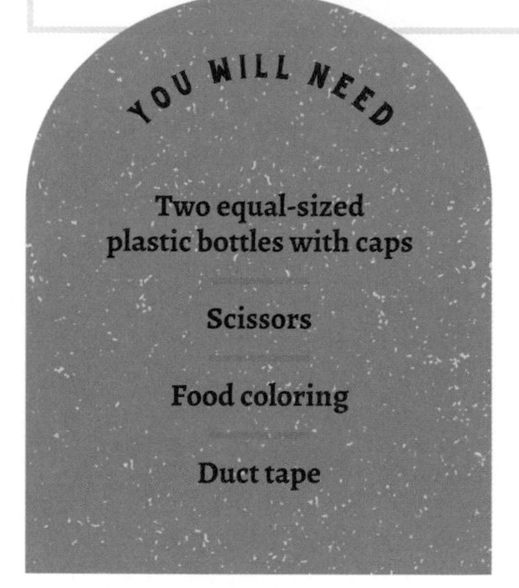

Remove the cap from one of the bottles. Using the scissors, cut a neat, circular hole in the middle of the cap. The hole should be about ½ in (1 cm) across. This is tricky, so you may need an adult's help. Now make an identical hole in the cap of the second bottle.

Add a couple of drops of food coloring to one of the empty bottles, then top it up with water. The water should almost reach the top of the bottle. Swirl the bottle around to make sure the food coloring and water mix together. The coloring makes it easier to see the tornado.

Screw the caps back on to both of the bottles as tightly as you can. Place the empty bottle upside down on top of the bottle that contains the water. Tightly wrap lots of duct tape around the lids. In a moment you're going to turn it all upside down and you don't want the water leaking out. This step is also tricky, so you may need to bother an adult again.

Real tornadoes form in a similar way. They occur when a downward current of air from a thundercloud sucks in air from its surroundings, creating a rapidly spinning column of air.

- Carefully turn the device upside down and stand it on a table. What happens? The bottom bottle may look empty, but it's actually full of air. The water in the top bottle is pressing down on the air, but the air in the bottom bottle is pushing back against the water. This keeps the water in the top bottle.

- Pick up the bottles and swirl them around in a circular motion. Now what happens? The movement allows air to escape upward into the top bottle. This creates a column of spinning water that is wider at the top and narrower at the bottom. This is your tornado. The water will now start to pour through the connection. Time how long the water takes to flow into the bottom bottle.

MAKE MUSHROOM PRINTS

Mushrooms aren't plants or animals. They are fungi. Try making beautiful mushroom prints with the fungi from your fridge.

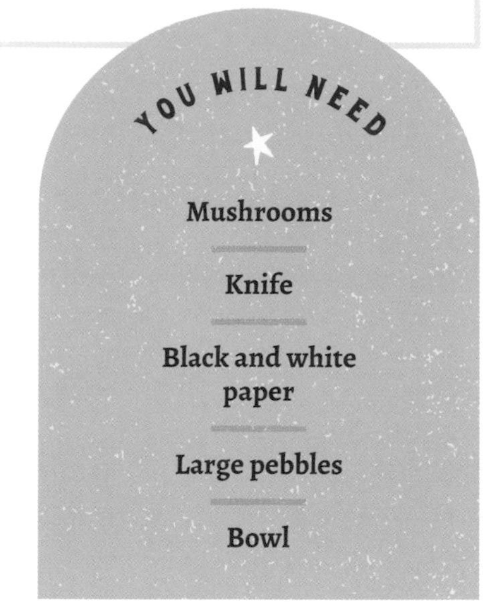

Collect some mushrooms. Some wild-grown mushrooms can be poisonous if they are eaten, so it's best to use store-bought mushrooms for this experiment. You're not going to eat them, but it's good to be safe. The mushrooms need to be fresh and juicy, not old and dried out.

The mushrooms that we see in the supermarket are the bit of the fungus that grows above ground. Let's start by learning about the mushroom's structure. Draw a picture of a mushroom in your science journal. Label the stem, which is the bottom part, and the cap, which is the top part. Look underneath the cap. Can you see lots of tiny folds? These are called gills, but they're not like the gills that fish have. Mushroom gills contain tiny seedlike structures called spores. When the spores are released from the cap of the mushroom, they can germinate and produce new fungi. Label the gills on your diagram.

It's time to make the print. Take two mushrooms that are the same. Cut off their stems and discard them. Place one cap on a piece of white paper, and the other on a piece of black paper. The gills should be facing down. Place a large pebble on each mushroom. The pebble should be big enough to weigh the mushrooms down, but not so big that the mushrooms are squashed.

Cover the mushrooms with a bowl to help prevent them from drying out. Leave the mushrooms overnight. In the morning, carefully remove the bowl, pebble, and mushroom. How do your prints look?

DID YOU KNOW?
Fungi can't eat food like animals, and they can't make their own food like plants. Instead, they absorb nutrients from nearby plant or animal matter.

The prints are made from the thousands of spores that have fallen from the gills. Different mushrooms have different-colored spores. White spores show up best on black paper, and darker spores show up better on white paper. What color are your spores?

Repeat the experiment, this time with a different variety of mushroom. Supermarkets sell many different kinds of mushrooms including oyster, portobello, and the more familiar white mushrooms. What do their spore prints look like? Remember that the prints contain live spores, so don't put them in your journal. Photograph your prints and stick the photos in your journal instead.

MAKE A CLOUD IN A JAR

The closest that most of us get to clouds is when we fly through them in an airplane. Make a mini cloud in a jar and learn how clouds form.

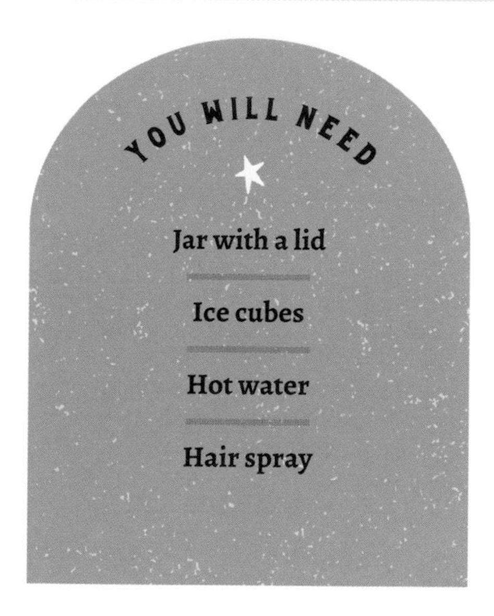

It is safest to do this experiment in the kitchen sink. Unscrew the lid from the jar and turn it upside down. Place three or four ice cubes inside the upturned lid. Ask an adult to help you fill the glass jar halfway with freshly boiled water, then balance the upturned lid on top of the jar. Watch the jar for five minutes. What happens?

Empty the jar and repeat the experiment. Fill the jar halfway with freshly boiled water, but this time, add a good squirt of hair spray into the jar before placing the cold lid on top. Watch for five minutes. What do you see this time?

Water evaporates from our oceans, lakes, and rivers, and rises up into the air as water vapor. As the water vapor gets higher, the air gets cooler. This makes the water vapor condense into tiny droplets of water. When droplets join together, they become heavier and fall as rain.

- In the first experiment you probably saw tiny drops of water form on the sides of the jar. The warm water vapor rose up inside the glass and condensed into droplets when it touched the cold lid.

- For clouds to form, something else needs to happen. The tiny droplets of water vapor in the sky need to meet and mix with tiny particles of dust, ice, or sea salt. The droplets and the particles stick together, and this makes a cloud.

- There were no dust particles in the first experiment, but in the second experiment, the hair spray acted like dust. Tiny water droplets stuck to particles of hair spray. This made the cloud in the jar.

- Now for the fun part. It's time to set your cloud free. Carefully remove the lid and watch the cloud disappear. If you've ever wanted to touch a cloud, now's your chance!

- See if you can make better clouds by squirting other harmless aerosols into the jar. Can you make clouds with deodorant or air freshener?

DID YOU KNOW?
Other planets have clouds, too, but they're not made of water. Jupiter has clouds that contain molecules made of ammonia and sulfur.

PREDICT THE WEATHER

Weather forecasters predict the weather using a device called a barometer. Make your own barometer and see if your weather predictions are better than the professional ones.

YOU WILL NEED

- Balloon and rubber band
- Scissors
- Glass jar
- Colored tape
- Paper straw
- Piece of card 8.5 × 11 in (21.5 × 28 cm)
- Ruler and pen

Our planet is surrounded by a layer of air that is nearly 60 miles (97 km) thick. This is called the atmosphere. The atmosphere presses down on earth creating atmospheric pressure. As the air warms and cools, and picks up or loses water, the atmospheric pressure changes. You can detect these changes with your barometer.

Cut the neck off the balloon and throw it away. Stretch the main part of the balloon over the opening of the jar. Pull it tight to get rid of any bumps or creases, then secure it in place using a rubber band. The air inside the jar is now unable to escape.

Put a small piece of tape over one end of the straw. Use the tape to attach the straw to the middle of the stretched-out balloon. The straw should now be horizontal and sticking out at right angles to the jar.

As the atmospheric pressure changes, and air presses down on the balloon more or less, the straw will move up and down. To see how much movement there is, you need to make a scale.

To make the scale, fold the piece of card in half lengthways. Using a ruler, draw horizontal lines spaced 1/2 in (1 cm) apart down one side of the card. On the top half of the card draw a sun. On the bottom half draw a cloud.

Set up your barometer indoors, in a place where the temperature doesn't change much. Don't put it near a window or a heating vent, as this will make the air inside the jar expand or contract, and affect the readings. Place the scale next to the jar so the straw is pointing at one of the lines.

Take readings every day. Does the dial move up or down? Make a note of your readings. How do they compare with the local weather forecast?

When the atmospheric pressure is low, there is less pressure pushing down on the balloon, and the dial points down. This means that rain is coming. When the atmospheric pressure is high, there is more pushing down on the balloon, and the dial will point up. This means it will be dry and sunny.

BUILD A THERMOMETER

Traditional thermometers measure the temperature using a liquid metal called mercury, but you can make your own thermometer using just the water from the faucet. Make a record of the daily temperature.

Fill the bottle with water almost to the top and add a few drops of food coloring. Swirl the bottle around so the food coloring mixes with the water.

Measure down 2 in (5 cm) from the top of the straw and make a mark. Measure down 4 in (10 cm) from the top of the straw and make a second mark. Roll a big lump of modeling clay or Plasticine into a fat sausage and wrap it around the straw to make a doughnut shape. Adjust the doughnut so the top of the ring is level with the second mark.

Place the straw into the bottle. Make sure the bottom of the straw doesn't touch the bottom of the bottle. The doughnut will be a lid and prevent the straw from falling into the water. Mix up a little more water and food dye, then use the pipette to drip some of the mixture into the straw. The aim is to raise the water level so that it reaches the first mark.

Add two drops of oil to the straw. Oil and water are immiscible. This means they don't mix, so the oil will form a layer on top of the water. This will stop the water from evaporating.

Your thermometer is ready to use. Stand it in a bowl of icy water. What happens to the water level? Mark the new water level with a pen. The temperature of icy water is 32°F (0°C), so this new mark also corresponds to 32°F (0°C). Write a "32" (or "0") next to the mark.

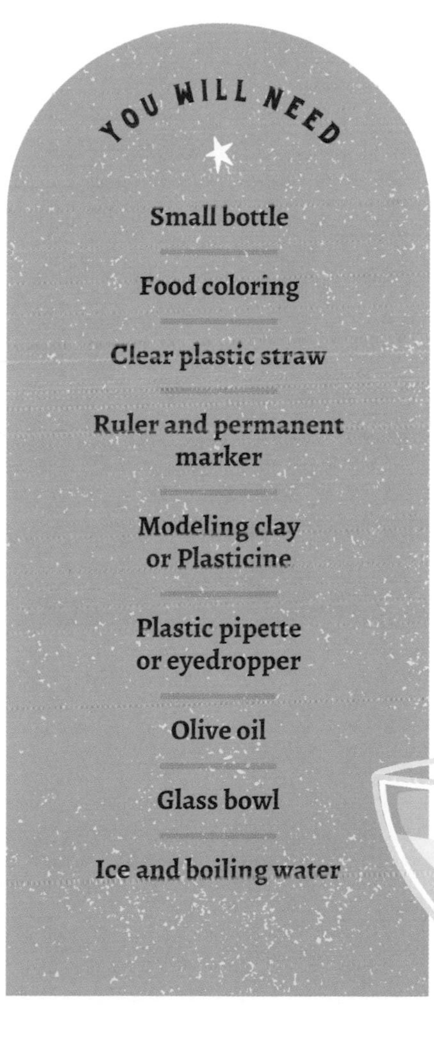

Try estimating other things. What's the temperature outside? How about your bathwater? Is it closer to 212°F (100°C) or closer to 32°F (0°C)?

Now stand the thermometer in a bowl of water that has just boiled. (Ask an adult to help you with this.) What happens to the water level now? This second new mark corresponds to the temperature of boiling water: 212°F (100°C). Write "212" (or "100") next to the mark.

WRITE WITH INVISIBLE INK

You don't have to be a spy to write in invisible ink. Learn how to create and decipher secret messages using just lemons and a candle.

- Cut the lemon in half and squeeze all of the lemon juice into a bowl. Remove any seeds that fall into the mixture. Add a couple of teaspoons of water and stir it well. The secret ink is ready to use.

DID YOU KNOW?
During the Second World War, prisoners of war wrote messages in invisible ink. They didn't have lemons or orange juice, so they wrote their secret messages in urine, which is also weakly acidic.

- Dip your paintbrush into the ink and write your secret message on a piece of paper. Work quickly because as the ink dries, it becomes impossible to see. If you have an old-fashioned calligraphy pen, you can dip the nib in the ink and write with that. Allow the paper to dry.

- Ask an adult to decipher the secret message. Get them to light a candle and hold the paper above the flame. Don't hold it too close or the message will go up in flames! Move the paper around over the heat source. What do you see? The message will be revealed.

This invisible ink works because the acidic lemon juice browns more quickly than the rest of the paper. What else could you use to make invisible ink? Try making invisible ink from other acids, such as vinegar, apple juice, and orange juice. Which works the best? Can you think of any other ways to reveal the secret message?

Heat is the key. Alternatively, place the paper on a heating vent or ask an adult to iron the paper for you. Both methods should work. Another way to read the message is to put salt on the drying ink. Leave the salt for one minute, then wipe it away. When you want to display the message, color over it with a wax crayon, and the message will appear.

YOU WILL NEED

A lemon

Knife

Bowl and water

Thin paintbrush

Paper

Candle and matches

BUILD A MINI RAFT

This is a great activity for a rainy day. Build a mini raft, then test it in the kitchen sink. How many coins can it hold before it capsizes?

Collect at least five long, straight sticks. Break off any side branches that are poking out and then snap the sticks so they are all roughly the same length. Your raft needs to be able to fit comfortably in the kitchen sink, so don't make the sticks too long or too big.

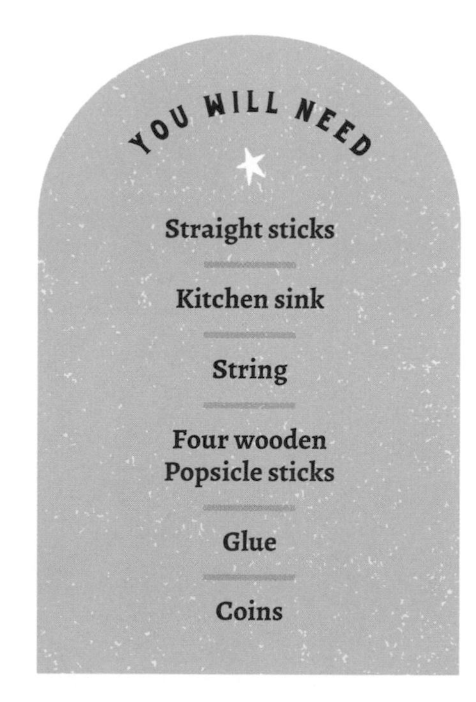

Lay the sticks out next to one another to form the base of the raft. Tie the sticks together using the string. The best way to do this is to tie the string around one end of the first stick and make a knot. Then loop the string around the end of the next stick and make a knot. Then loop the string around the end of the next stick . . . and so on, until you reach the last stick. Make a final knot to keep the last stick secure.

Now do the same for the other end of the raft. Secure the sticks by tying the ends together with string.

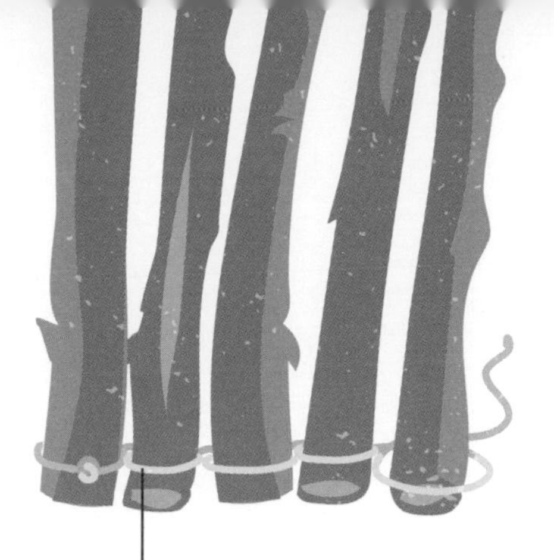
Knotting the raft

Rafts may look basic, but they are often used to move heavy items around. Timber rafts are sometimes used to transport enormous felled trees down rivers.

Lay branches horizontally and glue two Popsicle sticks on top of them facing vertically. Now turn the raft over and do the same on the other side. This will give the raft extra stability. Allow the glue to dry thoroughly.

Fill the sink with water and lower your raft in. Does it float or sink? Guess how many coins your raft will be able to hold before it capsizes. Write your prediction down in your science journal. Now add the coins one at a time. Were you right?

How can you alter the raft to make it hold more coins? Can you make it bigger? Take a look in the kitchen cupboards and around the yard. Is there anything you can add to make it more buoyant? Now repeat the coin test. Did you manage to improve your raft?

Balance coins on top
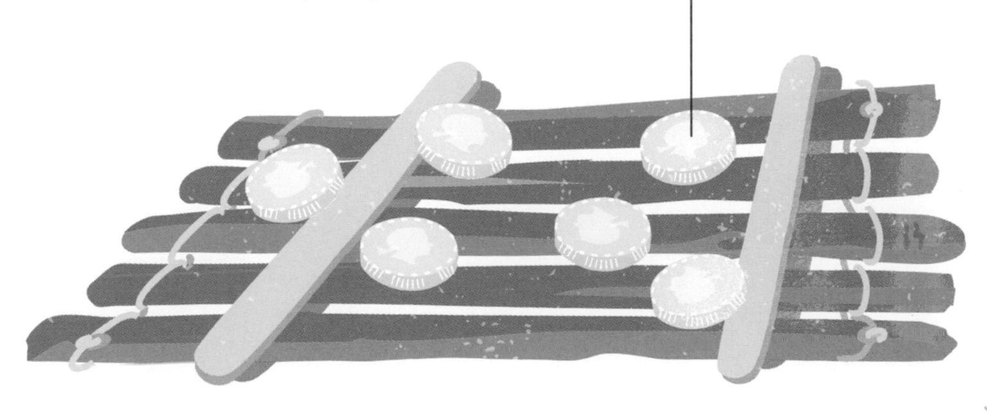

DIY FOSSILS

Why wait millions of years for fossils to form when you can make your own fossils today? Follow this recipe and make fossil imprints from the living things you find around you.

YOU WILL NEED

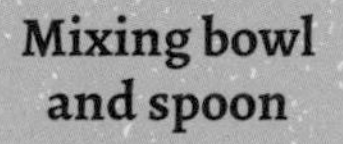

- Mixing bowl and spoon
- 3½ oz (100 g) plain flour
- 3½ oz (100 g) table salt
- 6 fl oz (175 ml) water
- Rolling pin
- Cookie cutters or a glass
- Items from outdoors, such as leaves and pine cones

Fossils are the remains of prehistoric life-forms that lived millions of years ago. Sometimes plants and animals become fossilized, but sometimes their footprints or outlines become etched into stone. These are called trace fossils, and they're very important because geologists can learn a lot from them.

To make some modern trace fossils, first prepare the dough. Add the flour and salt to the mixing bowl, and give it a stir with a spoon. Add the water, a little at a time, mixing as you go. The mixture should become doughy. When it's ready, it should come away from the side of the bowl cleanly.

Sprinkle a little flour on a work surface. Scoop up the dough with your hands and shape it into a ball on the powdered surface. Roll the mixture flat until it forms a layer around 3/4 in (2 cm) thick. Cut out circles using a cookie cutter or an inverted glass.

Choose an item for your first trace fossil. Flowers and leaves leave beautiful imprints, as do pine cones and empty snail shells. Gently press the item into the dough. It shouldn't disappear completely, but it should sink slightly into the dough. Press down on all of the features to make a good imprint.

Carefully remove the item from the dough using your fingers. What do you see? There should be a detailed impression of the item left in the dough. Place the dough somewhere warm, where it can dry overnight, near a heating vent or on a windowsill.

Look up some photos of fossilized plants, seeds, and shells on the Internet, like the picture above. Evolution causes living things to change over time. How similar or different are your fossils to ones from the distant past?

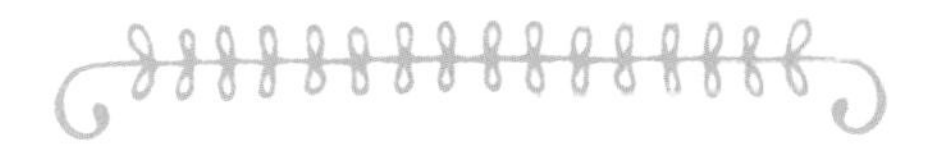

DID YOU KNOW?
Trace fossils include footprints left by prehistoric animals, but they also include other items that animals leave behind. These can include nests, burrows, and even dinosaur poop!

WASH WITH HORSE CHESTNUT SOAP

The seeds of the horse chestnut tree contain soapy chemicals called saponins. Make some horse chestnut soap and then see how well it works.

YOU WILL NEED

- About thirty fresh horse chestnuts
- Sharp knife and scissors
- Cheese grater
- Large bowl and water
- Cheesecloth
- Small plastic container
- Old piece of fabric
- Regular bar of soap

Peel the skin from freshly fallen horse chestnuts using a sharp knife. Horse chestnuts are quite small, so this can be tricky. You could ask an adult for help. Grate the horse chestnuts into tiny pieces using the cheese grater. This is even trickier. You'll almost certainly want to ask an adult for help. To avoid grating your fingers, hold each chestnut firmly and grate them slowly.

Fill the bowl up with warm tap water and drape the cheesecloth over the top. Press the cloth into the water so the middle part is submerged and the edges are hanging over the sides of the bowl.

Collect all the horse chestnut gratings and place them in the middle of the bowl, inside the cloth.

Leave the mixture to soak for a couple of hours, then gather the edges of the cloth into a bundle and remove it from the water. The horse chestnut pieces should still be inside the cheesecloth.

Squeeze the cloth to remove the excess water. Keep squeezing until the mixture inside the cloth forms a tight, compact lump. Open the cloth up and squish handfuls of the mixture into the plastic container. Leave the soap to dry in a warm, dry space overnight, then turn it out of the container.

DID YOU KNOW?
Many people think the Vikings were dirty and smelly, but historians think that they made horse chestnut soap and used it for washing.

It's time to test the soap. Take the old piece of fabric out in the yard and kick it around in the dirt. Make it as messy and dirty as possible, then cut it in half to make two equally dirty, smaller pieces of cloth.

Fill the sink with warm water. Scrub one piece of dirty cloth with regular soap, and the other with horse chestnut soap. Which soap cleans the best?

KEEP EXPERIMENTING!

So, did the experiments work? What have you learned? What was the best, the messiest, and the most unexpected finding? What worked well? What didn't go so well?

The great thing about experiments like these is that there are always more things you can try. In science, one question always leads to another. So if, for example, you dyed a T-shirt with the natural avocado dye (pages 154–55) and it turned out very pale, what could you do to make the color stronger?

This book may almost be finished, but the experiments don't have to be. Plan some more experiments of your own. Start by asking a question, like the one above, and then use that question to make a hypothesis. A hypothesis is just a sentence that describes what you think is likely to happen. For the avocado experiment, a new hypothesis could be: adding pits and skins to the mix will make the color stronger.

DID YOU KNOW?
More than 500 years ago, people were writing with avocado ink. Some of the ruby-colored documents exist to this day.

★ Now you can plan your experiment. Good experiments always have "experimental" and "control" groups. The control group is the part of the experiment where nothing changes. This gives you a benchmark against which to measure any changes that you make. The experimental group is the part of the experiment where you test new things.

★ So, dye one T-shirt using the original recipe (with pits only). This is the control group. And dye another T-shirt using your new recipe (with pits and skins). This is the experimental group. Then compare the results. Was your hypothesis correct? If the T-shirt in the experimental group was darker than the T-shirt in the control group, then your hypothesis was correct.

★ Now, whatever you do, please don't restrict yourself to the experiments in this book. Ask questions, make hypotheses, and design experiments about the world around you. Find out things that no one else knows, and if your hypothesis turns out to be wrong, don't worry. Scientists do sometimes find that their hypotheses are wrong, and this is how some of the most exciting breakthroughs in science are made. Scientists are continually revising what they know about the world as they perform more and more experiments to test their hypotheses. Be curious, be inventive, and be bold, but most of all, enjoy your garden science.

CREDITS

AUTHOR'S ACKNOWLEDGMENTS

Thanks to Kate Duffy, Lindsey Johns, and Sarah Skeate for their brilliant editorial and artistic skills. And thanks to my family and my dog. For being there.

PICTURE CREDITS

The publishers would like to thank the copyright owners for permission to reproduce their images. Every attempt has been made to obtain permission for use of the images from the copyright holders. However, if any errors or omissions have inadvertently occurred, the publishers will endeavor to correct these for future editions.

Alamy Stock Photos: Anne Elizabeth Mitchell 23; A. D. Fletcher 27; blickwinkel 57; daisyforster 97; dalekhelen 41; Dorling Kindersley Ltd 29, 43, 83; Linda Kennedy 49; Minden Pictures 89; Roger Parkes 53; George Philip 61; Adrian Sherratt 161; Dani Simmonds 99.

Carly Schmitt 167.

Getty Images/The Image Bank/Jeffrey Coolidge 109.

iStock: 13; 103.

Science Photo Library /Steve Gschmeissner 67T.

Shutterstock: 6, 7, 10, 11, 35, 45, 54, 67BR, 71, 88–89, 92, 93, 103, 107, 119, 125, 129, 137, 143, 144, 145, 148, 149, 163, 171, 175, 187, 189, 190.

Wikipedia Onderwijsgek at nl.wikipedia 66.